PENGUIN BOOKS

A DAY IN THE LIFE OF THE BRAIN

Baroness Susan Greenfield CBE is a Senior Research Fellow at Lincoln College, Oxford University. A scientist, writer, broadcaster and Cross-Bench member in the House of Lords, she has been the recipient of 32 honorary degrees from both British and foreign universities, and of many awards including Chevalier Legion d'Honneur from the French Government and an Honorary Fellowship from the Royal College of Physicians, as well as being selected as Honorary Australian of the Year in 2006.

SUSAN GREENFIELD

A Day in the Life of the Brain

The Neuroscience of Consciousness from Dawn Till Dusk

PENGUIN BOOKS

PENGUIN BOOKS

UK | USA | Canada | Ireland | Australia
India | New Zealand | South Africa

Penguin Books is part of the Penguin Random House group of companies
whose addresses can be found at global.penguinrandomhouse.com.

First published by Allen Lane 2016
Published in Penguin Books 2017
001

Grateful acknowledgement is given for permission
to reproduce or adapt the following:
Fig. 3 (Plate 3) from *Chemical Society Reviews*, 35, 897 (2006), reprinted
with the permission of the Royal Society of Chemistry; Fig. 4 (Plate 4) modified
from *European Journal of Neuroscience*, 32, 793 (2010), with the permission of
John Wiley and Sons; Fig. 8 from *Experimental Brain Research*, 182, 495 (2007),
reprinted with the permission of Springer; Fig. 9 modified from *Progress in
Brain Research*, 150, 12 (2005), with the permission of Elsevier; table on p. 215
modified from *Nature Reviews of Neuroscience*, 5, 713, Box 2 (2004),
with the permission of Nature Publishing Group.

Set in 9.35/12.31 pt Sabon LT Std
Typeset by Jouve (UK), Milton Keynes
Printed in Great Britain by Clays Ltd, St Ives plc

A CIP catalogue record for this book is available from the British Library

ISBN: 978–0–141–97634–1

www.greenpenguin.co.uk

Penguin Random House is committed to a
sustainable future for our business, our readers
and our planet. This book is made from Forest
Stewardship Council® certified paper.

In memory of Reg Greenfield, 1915–2011

'You can't use a knife of butter to cut butter.'

Contents

Preface

My father always liked taking things apart and was fascinated by how anything and everything around him worked, be it his car, a TV, a jet engine – or the body and the brain. Ever since I can remember, he would ponder in awe on the nature of electricity, human nature, belief and the precarious state of being alive, not with any fixed assumptions but revelling in the profundity and richness of the intellectual challenge of *not* having a simple solution. As James Thurber once remarked, 'It is better to know some of the questions than all of the answers,' and I think this sheer enjoyment of savouring a problem and sharing it with others must have rubbed off on me very early: it certainly set me on the path for my choices at school. Science was taught as a series of known facts, and no room was left for further wonderment: the amoeba split in two; distilling water (of which no one explained the relevance) involved drawing with stencils a conglomeration of equipment involving conical flasks connected by ruler-sharp tubes and required neatness in the exercise book but not much else; the physics of time and space was encapsulated by ticker-tape trolleys spewing out white strips of paper with appropriately distanced dots stamped out on them. Despite the changes in education in the digital era, I suspect there is still room in science teaching to show students the more distant and less certain horizons to which the bread-and-butter rites of passage might lead. In my case, the big questions that trouble us all (but especially, perhaps, adolescents) as to why wars start, the nature of love, free will, destiny and, above all, the essence of one's own individuality seemed to be so much better met by history and literature – in my case, of the ancient world. Accordingly, with great relief I abandoned science as soon as

I could in favour of the old Victorian package of Latin, Greek, ancient history and maths.

The world of the ancient Greeks, in particular, offered the chance to explore what seemed to be the big questions about the human condition, and inevitably led into a more general interest in philosophy. However, in my first year at Oxford, I was brought up short by an emphasis on the seemingly forensic analysis of language: I can still recall sitting in the Bodleian Library one Saturday morning, ploughing through a whole chapter on the definite article ('the') and reflecting that perhaps I had not, after all, made the best choice of subject. So it was that I changed to the then fledgling subject of psychology, grew increasingly to favour the more physiological options and became mesmerized, for the first time, by a science that didn't have simple answers but could address the very questions I'd asked as a schoolgirl with empirically based discoveries. To everyone's surprise, not least my own, I morphed into a neuroscientist, thanks to the huge support and encouragement of my then tutor Dr Jane Mellanby and the Head of Pharmacology, Professor William Paton. This was the start of a research career investigating novel brain mechanisms, with particular relevance to neurodegenerative disorders.

But that early fixation with the mind never left, and with it the nagging issue of consciousness: what it was and how it happened – which perhaps amounts to one and the same question. Yet, if someone claimed to be able to answer that question, what would I have expected them to show me? A performing rat? A brain scan? A formula? Not even the most hypothetical and futuristic of scenarios could capture that essential or, rather, the quintessential ingredient: subjectivity. So, alongside the daily lab experiments, I carried on talking to philosophers, in particular, the late Susan Hurley. Together, we organized a wide-ranging series of debates between philosophers and neuroscientists that ended up being published in 1987 as *Mindwaves*. The appeal of these broad discussions, which often lasted long into the evening, was that I learnt how a topic common to both disciplines – say, memory – could clearly be approached with a completely different agenda, set of priorities and perspective. Above all, for me as a neuroscientist, the most important issue was the risk of neglecting the phenomenology – the all-important subjectivity that caused one wag to label the neuroscientific exploration of consciousness as a CLM: a Career Limiting Move.

It was continuing reflections on the subjective nature of 'mind' and 'consciousness' that then led me, in 1995, to write *Journey to the Centres of the Mind*, and, in 2000, *The Private Life of the Brain*. Since you could 'lose your mind' but still be conscious, it struck me that the two terms were far from synonymous and that working out how they related to each other, in neuroscientific terms, might offer a small step forward. *The Private Life of the Brain*, therefore, is mainly a theoretical foray, albeit based on a range of on empirical data, into how 'mind' and 'consciousness' could be rooted in the physical brain and how we might develop a system of studying these often interlinked but sometimes independent phenomena by correlating objective events with their respective subjectivity. I ended up concluding that we need some kind of Rosetta Stone, a frame of reference that is bilingual, something that could be described in phenomenological terms just as readily as in physiological ones.

The perfect candidate was not any macro-scale brain region, nor any micro-scale collection of synapses, but a mid-level, 'meso-scale' brain process which, until the 1990s, had escaped detection: neuronal assemblies. Assemblies can be thought of as being a bit like the ripples generated by a stone thrown into a puddle: once triggered, large numbers (millions) of neurons generate a spread of activity, working together over a sub-second time frame. This fast scale of milliseconds had meant that assemblies were impossible to detect with classic brain imaging, which relies on indirect measurements, such as blood flow, and therefore typically has a resolution of only seconds. However, with the pioneering introduction of voltage-sensitive dyes which give a direct read-out of neuronal activity, it became possible at last to explore these highly transient coalitions of brain cells in real time.

When I wrote *The Private Life of the Brain*, the study of neuronal assemblies was still in its infancy, and I could only gesture at how they might provide the all-important basis for developing a neural correlate of consciousness: in fact, looking back at the latest edition, I see with some bemusement that, though mentioned in the text, the term 'assembly' doesn't even feature in the index. But I longed now to progress from theory to reality and to monitor assemblies in my own lab: fortunately, the usual obstacles of finding the right people and, of course, finding funds turned out on this occasion not to be as insuperable as I'd imagined. A talented graduate student in our group, Ed Mann, was willing to travel to Japan and was generously hosted in the lab of Dr Ichikawa for the summer in 2001 to learn the technique of optical imaging so that we could set up our own system back in Oxford. Just as importantly, and far from inevitable, funding for the highly specialized equipment and subsequent experiments was provided by an initial grant from Pfizer Ltd, and then continued by the Templeton Foundation, the Mind Science Foundation and the European Society of Anaesthesiology. What had seemed an improbable dream became a reality and, over the last fifteen years, we have been able to test out many of the ideas that were initially only hypothetical. Hence, this book.

A Day in the Life of the Brain is far from being an exhaustive review of the field of consciousness research, though it inevitably draws, where appropriate, on relevant and recent work in philosophy, psychology, neuroscience and physics. Sadly, it's unavoidable that things can get a bit technical, because, this time, unlike in *The Private Life of the Brain*, we'll be looking at many more real experiments. However, as much as possible, I've tried to save the general reader from too much scientific detail by separating out into the Notes the kind of information the specialist might expect.

The main aim of this book is to suggest a new interdisciplinary approach to consciousness: the central supposition is that neuronal assemblies offer a framework of description that is equally feasible using phenomenological and physiological terminology in parallel. Accordingly, the best possible examples of different subjective states of mind seemed to be not some contrived laboratory scenario but those with which we are all familiar: the different stages of a typical day. The plan is to journey through the ups and downs of waking, eating, working, playing, being troubled and dreaming to see how, in each instance, we can match up a particular subjective state with a differing profile of objectively measurable events in the physical brain: neuronal assemblies.

Since *A Day in the Life of the Brain* is heavily dependent on the papers we have compiled and/or had published by using optical imaging, I would like to thank those who have worked in the group on the project over the years and are co-authors on the publications cited. In particular, I would like to acknowledge the most recent researcher, Scott Badin, who has refined our study of assemblies to a much greater degree of precision and whose work features extensively throughout these pages. A special thanks is due to Dr Francesco Fermani, a theoretical physicist with whom Scott and I have passed many enjoyable evenings over a bottle of wine, attempting to go beyond empirical data to mathematical modelling, as described in the final chapter. Above all, I would like to thank Dr Ian Devonshire: not only did Ian progress from working with brain slices to performing optical imaging *in vivo*, but he has now made an invaluable contribution to the book by fact-checking, editing and ensuring that the references are up to date and appropriately representative.

The book itself would not have been possible without the support of the editors from Penguin, Stefan McGrath, who originally commissioned it, and Laura Stickney, who has tirelessly seen drafts through various iterations into a form ready for the final excellent copy editing of Sarah Day. Lastly, as always, I'd like to thank my agent, Caroline Michel, for her unerring enthusiasm and friendship, and to acknowledge the contribution of others, which has been less specific but no less valuable: Professor John Stein, Professor Clive Coen and Mr Charlie Morgan have provided endless intellectual challenge and tough love, while my mother, Dorice, now an author herself, has provided the best input of all: unconditional love. So we come full circle back to my father, who gave both my brother and me a real curiosity and courage to square up to the big questions. I'd like to think that, wherever he is, he knows how much I owe him.

Susan Greenfield
Oxford, 27 March 2016

1
In the Dark

THE STORY SO FAR

It's early in the morning and still dark outside. Your heart rate has dropped to a single beat per second and your blood pressure is the lowest it will be for the next fifteen hours. Your respiration has fallen to twelve breaths a minute, while your blood glucose is at rock bottom. Your bladder and bowels are slowly filling but are not yet sufficiently inflated to disturb your slumber. Despite the ingenious physiology remorselessly at work in all your organs, you sleep on, oblivious. In this state, your brain and, indeed, your whole body are currently somehow preventing entry to a private inner universe that poses one of the biggest challenges in science – but any minute now, when the alarm buzzes, this special place will be all yours, exclusively. However close you are to someone, however articulate, poetic, musical or compassionate you are, you will never, ever, be able to share your direct subjective experience with anyone else. But right now, for just a little longer, you're 'dead to the world': unconscious.

Without consciousness, life would indeed be pretty much the same as death. The conscious condition makes life worth living: yet what *is* it, this insubstantial, intangible inner . . . *what*, exactly? Across the centuries, our predecessors have struggled to define and understand how we can best grapple with what is so intellectually elusive yet so familiar that we take it for granted day after day. Over the last forty or fifty years, the issue has been thrown into ever sharper focus with the ascendance of neuroscience and the massive increase in our knowledge of the brain. But with this wealth of facts and insight the problem becomes ever more glaring: how can your individual, subjective experience be translated within your physical brain into squirts of chemicals and electrical blips, and vice versa? When it comes to other ambitious scientific scenarios such as time travel or perpetual-motion machines, the possibility of their eventual realization might defy the laws of physics, but then again – at least, hypothetically – you'd recognize the solution as such if ever some ingenious inventor presented it to you. So what would actually prove to you that some scientist, or philosopher, or even sci-fi novelist, had come up with the definitive insight into the subjective experience we call consciousness?

Would the newly enlightened individual perhaps be brandishing a brain scan, or clapping their hands at a mathematical formula? These 'solutions', however ingenious, would never be convincing, as they would not account for *how* objectively observable events transubstantiate into the first-hand feeling of that very special private experience. Somehow, in some way, in the brain and in the body, a subjective perspective is generated – we could almost say 'conjured up' – but no one has so far been able to produce any kind of convincing answer, however hypothetical, as to how this apparent miracle takes place. It is one of the deepest mysteries and the most enduring puzzles not just for scientists but for any of us to ponder. Small wonder that the black night outside your bedroom window is as nothing compared to the intellectual darkness that gives rise to this conceptual impasse.

Back in the mid-1980s, the philosopher Susan Hurley and I thought it would be a great idea to set up a series of seminars at Oxford University between neuroscientists and philosophers on various aspects of the mind and brain. Inevitably, we made little progress in understanding what any brain-based process might be for consciousness itself,[1] not least perhaps because neither neuroscience nor philosophy has any clear common base from which to start, as the two disciplines share no opening assumptions. Nonetheless, the seminars did highlight the major headache that bedevils both disciplines: how to explore a subjective phenomenon in an objective way. One plus, however, was that we discovered that a fruitful multidisciplinary debate could be had over more modest, more specific questions, such as whether a machine could be conscious, what the evolutionary power of consciousness could be, and the impact of language on thinking.

Even back then, neuroscience was starting to challenge the tired old dichotomies of brain versus mind, mental versus physical, by showing how changes in the brain due, say, to physical damage, matched up, correlated, with changes in subjective personal experience. In the words of one of the philosopher participants, Paul Seabright: 'Memories, no less than molecules, are physical objects, part of the physical universe, but objects we are capable of recognizing under names other than [those] physics gives them.'

This tension between the objective and the subjective has seemed insoluble ever since the great philosopher René Descartes first raised it in the seventeenth century, when he chose to segregate a conscious mind from the biological brain. This conceptual chasm, between the first-person perspective of moment-to-moment experience in real life and the third-person perspective that is the hallmark of scientific experimentation, is tough to cross. When we describe a human being, or any animal, as having 'perceptual experiences' or 'conscious states', we are talking about some of the most basic features which distinguish living from non-living things. A stone is not 'about' anything and neither does it have any subjective properties, but a perception is 'about' what is seen (or experienced) and therefore does have subjective properties, such as what it is like to see the colour red. These features are obvious to us from our first-person view of the world, but when we shift perspective to the third-person stance of objective science we find ourselves in some difficulty. We can speak of brain states, which in turn may be reduced to neurons sparking and pulsing out powerful chemicals, but it is hard to see how this neurobiological maelstrom relates to the features of consciousness that are so apparent to us in your own everyday experience: the taste of chocolate melting slowly in your mouth, the sunshine on your face, the swishing roar of surf. The aim of the journey we're about to undertake in this book is to gain insight into what is physically happening inside our heads when we savour food, become tipsy after a couple of glasses of wine, dream, walk outdoors, or sit all day long in an office.

At each stage of your day, various factors drive an endless succession of different types of brain states, which are themselves based on very large numbers of brain cells working together for just a fraction of a second. 'Assemblies' are powerful brain processes which, nonetheless, are still relatively unknown and under-investigated. We will see how these transient, large-scale neuronal coalitions, assemblies, can correspond to your experience of different levels and types of consciousness, and so operate as a kind of Rosetta Stone – that famous relic engraved with a decree written in three languages and so providing the key to unlocking Egyptian hieroglyphs. By matching up these novel events in the brain with different types of consciousness, assemblies could serve a similar purpose, enabling us to translate one 'language', the subjectivity of personal experience, into another – the objectivity of neuroscience, and the other way around. After all, surely any so-called 'scientific' explanation of consciousness must give equal weight to the first-person subjective experience.

The basic assumption that subjectivity is core to understanding consciousness is something the scientific community has found a bit unpalatable to take on board because the empirical method is ruthlessly objective and impartial. No surprise, then, that someone once described the study of consciousness as a CLM: a Career Limiting Move; to many traditionalists, the spectre of being 'unscientific' prohibits any further exploration of the subject. 'Scientific' is an adjective frequently appended with much fanfare to all types of terms, especially 'evidence', yet it is rarely defined. Suffice it to say that scientists typically report findings which, most importantly of all, are objective: two or more researchers will each measure something in the same way under the same conditions and come up with the same, consistent result. Yet that's precisely why the normal machinery of scientific method is so tricky when we are dealing with consciousness, because the would-be researcher really has to acknowledge upfront the quintessential feature of subjectivity. Even those neuroscientists daring to work in the area – pleasingly, a swelling minority – tend to ignore the diversity of individual subjective states: instead, we focus, understandably, on principles of brain function with which we are already familiar and the ensuing processes and effects that we are comfortable measuring.

DISENTANGLING DIFFERENT TERMS

Even before you take your lab coat down from its peg, however, the problem of defining consciousness may stay your hand. What exactly are we going to be dealing with? Trading the term in for apparent synonyms such as 'wakefulness' or 'awareness' will get us nowhere; nor, in any case, are these alternatives satisfactory, as they emphasize a more generic, passive reactivity to the outside world rather than our quintessentially individual and unique appreciation of the ongoing experience of it. The difficulty in clarifying what we mean arises from the adoption of one of the easiest strategies we have for defining things. For instance, you might say, 'Flight is when you defy gravity'; but consciousness is when you do *what*, exactly? You don't necessarily have to do anything at all: you could simply lie flat in silence and with your eyes closed, yet you could still be conscious. This kind of *operational* definition is useless. The alternative option is to refer to a higher 'set', or wider category. For example: a table is a piece of furniture; love is an emotion; consciousness is a . . . is a *what?* Where is the higher, or wider, category? There is none.

Since we can't formally define consciousness, the next best thing is at least to sort out understandable confusions with other terms that can sometimes sneakily creep in as further synonyms when, in fact, they are concepts that have their own distinct meaning. For example, the subconscious, so central to the ideas of the father of psychoanalysis, Sigmund Freud, implies that there is a scenario of which you are not aware even though you are fully awake. The subconscious *could* be a necessary, but not sufficient, component of the final conscious state, as opposed to unconsciousness, its complete abolition in deep sleep or coma. Then there is self-consciousness, the sense of being a unique individual, which at full throttle might involve a feeling of ill ease, even anxiety. Self-consciousness in its more muted, everyday sense of merely being aware that you are different from everyone else must be in some way different from general consciousness: otherwise, why the need for two terms? And just think for a moment about non-humans: a cat or dog would, by every devoted owner's account, be decidedly and irrefutably conscious. Then again, it seems less plausible that the pet would be fully cognisant of their respective feline or canine station in life: after all, not even human infants are aware of who and where they are, yet we assume they can be conscious.

Nonetheless, many scientists and philosophers focus their efforts to understand consciousness on what is often referred to as a 'higher' state, or metarepresentation: they explore the extent to which the beginnings of self-consciousness could be apparent in non-humans such as other primates.[2] But a subjective state of consciousness will not guarantee at all that self-consciousness will automatically ensue. For instance, if you are in a state of literal abandonment – say, by means of wine, women and song, or the contemporary equivalent of drugs, sex and rock 'n' roll – then, almost by definition, it would be impossible to be self-conscious (or vice versa), as any reluctant would-be dancer knows when venturing up on to the floor. However, *self*-consciousness, fascinating, disconcerting and occasionally embarrassing though it may be, isn't really getting to the heart of the most basic problem – what it means, in neuroscientific terms, just to *feel* something: to be sentient. The tail-wagging dog, the purring cat and the gurgling baby are having some kind of subjective experience, but they are not self-conscious, although to all intents and purposes they behave as though they are conscious and we interact with them on that assumption. The crucial issue we need to sort out before anything else is the pure subjectivity of sensation – any old sensation – in the first place. The all-important first step is not necessarily to recognize yourself in a mirror, to know your own name, or be appreciative of your life story so far but, much more basically, to experience some raw kind of inner state.

But perhaps the most common and misleading muddling is when the term 'consciousness' is used interchangeably with 'mind'. When it is operational and in good working order, this 'mind' is something highly personal, hence expressions like 'my mind's made up', or 'to my mind', or 'change my mind'. But when you do lose consciousness, say, in sleep or anaesthesia, no one could describe your condition as one in which you had lost your mind. And just as this 'mind' will presumably endure during sleep, when consciousness is 'lost', so the converse will also hold: in moments of great pleasure, you can 'lose' your mind without losing consciousness – the wine, women and song scenario, in which you will 'be out of' your mind yet still be conscious, often 'ecstatically'; the very term comes from the Greek for 'to stand outside of yourself'. However, even when you experience extremes of emotion, you will still be experiencing an intensely subjective, conscious state. So whatever it is, the mind must be something continuously available yet something you need not necessarily be accessing all the time. The mind can be viewed as something that is independent of consciousness when you 'let yourself go', but at the same time it must be a way of describing some peculiarly personal features of your individual brain, things which make you the unique person you are and which will in turn contribute as a factor to many, but not all, moments of your consciousness.

Having cleared away, if not cleared up, some of these deceptive related terms, we're still faced with the elusive reality of this raw, primary consciousness. It might be that the best we can do in terms of a definition is to adopt what will be at least the most pragmatic way forward for a brain scientist – yet which may well make a philosopher wince – and say, 'Look, we all know what we mean by 'consciousness'. It is what people lose when they go to sleep, it is what is abolished when they are anaesthetized; it is the experience that you are having right now: an inner, subjective state that no one else can share.' We therefore have a choice: either to abandon ship right now because we cannot articulate a formal definition or, alternatively, to go along with this informal appeal to common sense. If we take, as we will, this more practical stance, then we can explore just how far neuroscientists can go to try to work out how it is all happening: how this mysterious process of subjectivity unfolds in the physical brain and body.

There are two possible ways to set about this. First, we could start with the actual physical brain and, subsequently, try to deduce some theory, or model, of consciousness from whatever physiological processes we observe. Alternatively, we could go with a second option and begin the other way around: first, to develop a theory, or model, of consciousness and, subsequently, apply this hypothetical scenario to see if it could work in elucidating the real brain. Let's try each strategy in turn: Experiment to Theory, then Theory to Experiment.

FROM EXPERIMENT TO THEORY

When starting with the physical brain itself, the huge challenge is to work out how consciousness might ever arise from the unlikely and unhelpful biological squalor of its morass of molecules, cells and electrical blips. In other words, we need to identify what scientists call the neural correlate of consciousness (NCC), that is, the single, all-important critical feature in the actual brain that matches up faithfully with a personalized and direct subjective experience. Candidate NCCs can come, as we're about to see, in a wildly diverse variety of flavours, from domineering brain regions and oscillating neuronal circuits down to single cells and tiny elements within them. But they all have a crucial, basic principle in common: all NCCs start by focusing on a particular brain feature which has, for whatever reason, been seized upon as particularly special. The game plan is then to attempt to link this special feature to consciousness as typically monitored by a laboratory or clinical protocol, for instance, by determining whether something is visible versus invisible, or the subject is conscious versus unconscious.

Many distinguished scientists have tried to solve the puzzle of consciousness by using this strategy. Perhaps the most notable was the late Francis Crick, who famously cracked the double-helix code of DNA in 1953 and several decades later turned his attention to the biology of consciousness. Crick's aim was to work out what in the brain is different when you see something, in contrast to when you are oblivious to it. His reasoning was that it should be possible to detect differences in the brain according to whether or not the human subject *reports* that they are conscious of something. When the individual affirms that something is indeed visible to them, then whatever new process we observe occurring in the brain will, with straightforward logic, be the physical correlate of consciousness. Consequently, Crick focused specifically on visual experiences not least, presumably, because they are easy to manipulate experimentally – so long as the subject keeps their eyes open.

The rationale for paring down the complexity of human consciousness to its bare essentials is clear: by focusing on just one of the five senses, in this case, vision, Crick and his colleague Christof Koch could keep things as simple as possible. Their idea was to demonstrate the key brain mechanism for consciousness as 'the minimal set of neuronal events that gives rise to a specific aspect of conscious percept'. In other words, the goal was to identify the bare minimum of brain processes that would measure up with, and accommodate, a conscious experience.[3]

These minimal neuronal events were then boiled down further within the visual system to a particular group of neurons 'associated with an object or event', which were constant and predetermined entities in and of themselves: a fixed feature of the central nervous system. This set-up immediately presents a problem: how can a pre-organized set of brain cells reflect a fleeting moment of consciousness? Crick and Koch's answer: 'To reach consciousness, some neural activity for that feature has to cross a threshold.' But the real problem there is that the crucial neural activity was never specified, and the only way a rigid network of cells could ever do anything above what it is already doing would be to become more active and discharge more electrical blips (action potentials). But, if this is the case, the possible repertoire of the all-important special brain property for consciousness is going to be severely limited, to a mere increase in common-or-garden cell activity. What would make that shift so vitally significant, some kind of neuronal Rubicon? We have no indication what the 'threshold' would be, or why, nor why a *quantitative* change in collective activity would be in any way so distinctive that it triggered a *qualitatively* different mental state.

But could we in any case really best understand a state of consciousness by boiling it down simply to one of the senses, to the exclusion of all the others? In a highly contrived experimental scenario in a lab, it is just about conceivable that a subject could concentrate for a short, predetermined period on a purely visual stimulus, but even then they would still hear the voice of the experimenter, still be aware of the hard or soft surface of the chair on which they are sitting, still detect perhaps its slight smell of leather or polished wood. In other words, the subject will be having a multisensory, holistic experience that isn't compartmentalized into five different types of consciousness corresponding to each of the five senses.

Moreover, Crick and Koch were at pains to stress that in the attempt to keep consciousness or, rather, its correlate, down to the minimum kit, they will 'leave on one side some of the more difficult aspects, such as emotion and self-consciousness'. However, our feelings and emotions might well be the all-defining quintessence of consciousness in its most basic form, as arguably demonstrated by the purring cat or the gurgling baby. If consciousness is to be simplified at all to a minimum kit, then surely it should be boiled down to a raw emotion that is the output of a holistic brain state – a shriek, or a cry, or a giggle – not to a single, isolated visual sense, say, floating free and independent of all else.

However, putting this objection aside, comparing scenarios in which you see something versus those in which you don't, could still get us somewhere: the idea now would be to investigate the particular conscious states of patients with brain damage as they react to stimuli presented to them under various conditions. By comparing the particular kind of impairment they have experienced with any deviations in their responses from those of the norm, scientists could extrapolate which brain regions have most influence over consciousness. For example, patients with damage to the fibres that connect the two cerebral hemispheres present with what is known as a 'split-brain': as the term suggests, the two hemispheres of the brain are 'split' from one another and unable to communicate directly. By showing, say, an apple to the right hemisphere (by placing the object to the left), the patient will be unable to name it. This is because many aspects of speech are linked to the left hemisphere in most people so, despite being able to see the apple, the patient cannot actually articulate what it is. When the apple is positioned to the patient's right, its image is projected to the left hemisphere, and the patient this time can come up with the right name.[4]

Another group of neurological patients who have proved popular with neuroscientists are those with 'blindsight', a condition in which they have a blind spot in their visual field so that they are not aware of objects positioned in that field of view.[5] Sometimes the blind spot can extend throughout their visual field to render the patient completely blind and yet, bizarrely, psychological tests reveal that the brain has nonetheless registered and processed objects. Those affected by the condition, for instance, will often be able to catch a ball thrown to them, despite not consciously being able to see either the ball or the person throwing it, almost as though they were on automatic pilot.

But there is a big 'but'. The problem with studying brain-damaged patients is that, although an investigation of these conditions may give insights into the bizarre type of consciousness a particular person is experiencing – its quality, or content – it tells you nothing at all about how that consciousness is generated in the first place. After all, the subject will have been conscious throughout the entire experiment.

Then again, it is possible to work with healthy subjects, this time by simply looking at shifts in consciousness. Some neuroscientists have even gone so far as to treat consciousness as pretty much synonymous with attention:[6] but they are not the same thing. For example, you can have attention without consciousness.[7] When a photo is flashed up briefly and unexpectedly, people can still summarize its content, even though they don't have time to register the detail:[8] in fact, it seems that thirty thousandths of a second is all that is needed to appreciate the 'gist' of a scene, without awareness of any of the specific detail it contains. And then you can have consciousness without attention. Even if someone is attending to some other task, they can still distinguish features of a peripheral scene, such as whether it does or does not contain an animal or vehicle[9] or whether a face is male or female.[10]

Intrinsically sound and inherently interesting though all these various findings might turn out to be, they still fail to throw light on the basic issue: understanding the actual transformation from unconscious states (anaesthetized or asleep) to having that tantalizing subjective inner experience. The deal-breaker surely is that split-brain and blindsight patients (or indeed, healthy volunteers in whom these conditions are effectively induced experimentally) are *continuously conscious*: the key, underlying phenomenon we want to explore remains mockingly constant. It's important to remember that the intransitive verb 'to be conscious' is not interchangeable with its transitive counterpart, 'to be conscious *of* something'.

Nonetheless, the strategy of comparing events in the brain which could match up with moments when you see something, in contrast to when you do not, has continued in popularity, with experiments in healthy individuals as well as brain-damaged patients, now that brain imaging with scans has become mainstream. For example, only during 'conscious' visual experiences of a particular object do the later stages of the visual pathway in the brain light up in brain imaging.[11] But, in the end, how much can really be deduced from knowing that particular brain regions, when lighting up in a scan, match up with a specific conscious experience? After all, a correlation can be a cause, an effect, or neither. The actual relationship between an active area of brain and a subjective experience is far from obvious and continues to defy explanation.

Anyone who looks in a continental butcher's shop window can see straightaway that the brains of different mammals are different in size, shape and appearance but are nevertheless composed of different regions readily visible to the naked eye and adhering to the same basic anatomical theme. So could any of these distinct brain regions qualify as 'a centre' in the brain for consciousness itself? After all, it's a happy thought that the brain is composed of autonomous 'centres' for this or that function, since the tenets of brain operations would then be so much easier to grasp. It was for this very reason, at the beginning of the nineteenth century, that the 'science' of phrenology (Greek for 'study of the mind') became popular. White porcelain heads covered with black-lined rectangles with precise labels such as 'love of country' and 'love of children' provided the template against which bumps on an individual brain could be evaluated in order to gauge the strength of a particular trait.

Nowadays, neuroscientists are all too well aware that this is a seductively simplistic concept and that there is in fact no such thing as a 'centre' for any mental function, let alone consciousness. There are two good reasons for rejecting any such notion. First, it just doesn't make sense to say that the brain is composed of independent mini-brains. When I was a child, I used to read a comic strip, *The Numskulls*,[12] about little men who lived inside a man's head; each had a job to do. There was the nose-cleaning department, for example, as well as some kind of HQ where the manager Numskull telephoned orders to his subordinates. I can still visualize well how the walls within the cartoon brain separating the different Numskull rooms were very rigid, the doors between them closed. In a sense, this is a rather crude caricature of how some people, when they talk about 'centres' in the brain, might be inadvertently conceptualizing consciousness. Just think about it: if this was how the brain truly worked, we would need to know what is going on inside each Numskull skull. Might there be mini-Numskulls and, inside the mini-Numskulls, micro-Numskulls, and so on? Surely, if we follow this strategy, we are merely going to miniaturize the problem, not solve it.

In any case, we now know from half a century or more of research that the brain doesn't work like that. Take vision, for example. There are at least thirty different brain regions that contribute to the experience of seeing something.[13] It's a bit like instruments in an orchestra, or a *cordon bleu* recipe with very complex ingredients. Each brain region does indeed have its own specific part to play, but each contributes to the whole. That whole is more than the sum of its parts and that whole is the moment of conscious experience.

Even so, one enduring article of faith is that consciousness must be related in some way to the cortex, the outer layer of the brain (so named after the Latin for 'bark'), since this part of the brain has been the latest to develop in evolution and is ever more pronounced in species of increasingly sophisticated mental prowess. Moreover, it appears that simple deactivation of the cortex is sufficient to create loss of consciousness.[14] So if that's all there were to it, then we could simply say that all the consciousness condition requires is an intact cortex that interfaces operationally with the rest of the brain. Such an idea – namely, that it's not, after all, the cortex itself that's all-important but the brain must additionally be intact – is too vague to be helpful.

A further reason for not putting all our eggs into the cortex basket is one arising from investigations that took place almost seven decades ago. The pioneering Canadian neurosurgeon Wilder Penfield suggested that the cortex may not be at all important for the experience of consciousness. This heretical proposal was based on the observation that there was no impairment in the continuity of consciousness of 750 awake patients undergoing surgical procedures involving removal of portions of the cortex.[15] In addition, much more recently, the idea that the cortex isn't required for consciousness has been further supported by investigation into cases of a condition known as anencephaly/hydranencephaly, in which babies are born with large portions of the brain, including the cortex, absent. Children with these conditions still show signs of wakefulness and consciousness, as assessed by standard neurological examinations, during their development and on into maturity.[16] Finally, within a perfectly intact cortex, loss of consciousness can still occur with 'absence' seizures.[17] In other words, the cortex cannot be the seat of consciousness.

Another flaw in any project of trying to localize consciousness in specific brain regions is apparent when you look at brain-imaging data monitoring the effects of anaesthesia. It turns out that there are many regions involved in the first-hand experience a subject may have at any one moment. Imagine there was indeed just a single and independent centre for consciousness: if so, you would predict that, if you were to anaesthetize someone and thereby take away that consciousness, then only that one special area, the 'centre' alone, would conspicuously and exclusively be rendered inactive. But that's not the case: imaging studies with human volunteers show that there is no single brain region in isolation that is out of action when someone undergoes anaesthesia; instead, the brain shuts down activity across the board.[18] There is no one area that is exclusively diminished in operations as the anaesthetic starts to take effect.

Inevitably, the highly improbable phrenological link between bumps in skull bone and subtle mental traits became discredited as advances in medical science made possible more direct, invasive examination of the real brain itself. Yet the simplistic reasoning of phrenology can still have echoes in subsequent clinical interpretations. As medicine progressed, clinicians became increasingly adept at keeping patients alive even if they had sustained serious brain damage from, say, bullets, or trauma, or stroke, leading in turn to particular neurological syndromes sadly still common today. But a misguided phrenology perspective still sneaked in from time to time: it remained very tempting to assign the damaged brain area the 'function' that had been lost. The flaw in this reasoning was highlighted by one psychologist wag over fifty years ago: if you remove a valve from a wireless (as it was then called) aka a radio, and the device started to howl, then you wouldn't claim that the function of the valve was to inhibit howling.[19] Of course, if the brain area in question is faulty, like the elderly valve, the holistic system of the wireless will be impaired, but the specific contribution of the valve, or indeed the brain region, cannot be extrapolated backwards from the net outcome.[20]

Despite these advances in understanding, an enthusiasm for the potential significance of different brain regions has been remorselessly sustained in neuroscientific research. We've seen that in the earliest investigations some thirty years ago cases of impaired functioning gave clues as to what brain area was linked to a specific problem. Now, however, healthy volunteers can be used to see what area of the brain is normally operational during certain behaviours, thanks to the huge leaps forward since the 1980s in non-invasive brain-imaging techniques. Nowadays, you can look at a scan and you might well see bright blobs pinpointing certain areas in a sea of background grey brain, or perhaps multicoloured arrays in which white is a hot spot, shading through to yellow to orange to red, to a low-activity purple perimeter. When confronted with these impressive and beautiful images, it's important to bear in mind that these pictures are opening up to you a window on to the brain at work *but* at work over a protracted period in time and with much of the vitally important ongoing brain activity missing from the picture.

The real issue arises over interpretation. Brain scans usually have a temporal resolution of seconds, while the action potential, the blip that is the universal electrical signature of active brain cells, is over a thousand times or so faster. Think of brain scans, therefore, as comparable to those sepia Victorian photographs showing in precise detail static, inanimate buildings but at the same time excluding any people or animals, which would be moving too fast for the lengthy exposure to capture. The buildings were, of course, present and real, as would be a steady-state tumour showing up on a brain scan, or a permanent lesion after a stroke; but they would constitute, literally, only a small part of the complete picture.

Nonetheless some scientists,[21] undeterred, have in their studies claimed that it is possible to 'decode' mental states from looking at brain imaging if they focus on the spatial pattern of responses across the cortex. However, the very term 'decoding' is misleading. A code, by its very essence, is a format that will always need to be translated back into the original and which must be very different from the original for it to make any sense. Dots and dashes, for instance, are meaningless unless you know Morse code; only then can decoding back into the original version take place. In contrast, subjective mental states, if they are indeed analogous to dots and dashes, are not converted back into spatially distributed brain activity, or vice versa – instead, they are correlates of it. In brain science, there isn't even as yet any framework for a causal relationship, let alone a rule book for code breaking to enable switching between the two.[22] Just because you know *where* something might be happening, it doesn't tell you anything about *how* it is happening – not least because it's far from obvious what the 'something' is, especially when captured over time windows two or three orders of magnitude slower than real time.

The most problematic issue with brain regions being the 'centre for' this or that mental property is as basic as the terminology itself. A 'centre' suggests that everything starts or finishes within it, yet there is no brain area that works in such a way. Instead, all brain regions interconnect with each other, even over long neural distances, via fibre tracts which enable constant dialogue across the brain, like a telephone exchange. There is no hierarchy of command within an individual, unlike the iconic Russian dolls each of which fits inside another. Somehow, each brain area is a kind of relay station – a junction, or a crossroads – but never, ever the final destination.

Another difficulty when it comes to brain scans is that we normally think of a mental trait, let's say being witty or kind, not in a *quantitative* way: how often we do it, or how much we have of it – but rather as a *quality* someone either happens to have or not have, and which makes up part of their individual character. Such refined and sophisticated traits arise mysteriously from the machinations and interplay of many different components of the holistic brain working from the level of individual synapses and proteins through to the sophisticated chatter of large-scale neural networks. As such, they defy simple operational definitions and are far harder to highlight on a brain scan than a quantifiable skill would be: phenomena as abstracted as wit and kindness are not readily measurable – unlike, say, memory capacity.

This doesn't mean, of course, that we are not making important new discoveries about brain function by using these scans: far from it. It's just that we shouldn't jump to easy conclusions. Take the well-known study of registered London taxi drivers, who famously have to pass an oral exam testing their knowledge of all the streets and one-way systems of the UK capital. Brain imaging revealed that the hippocampus, a part of the brain related to memory, is enlarged in these drivers, due to the huge burden on their working memory.[23] But that doesn't mean we can then leap to the conclusion that the hippocampus is the single 'centre' for the task in question: in this case, the complex and multifaceted process of memory. Memory involves the laying down of a diversity of new skills, new facts and new events, as well as their subsequent retrieval; all these different types of memory, and their different stages in processing over time, involve many brain regions and mechanisms.[24]

More recent attempts to find the key brain property serving as the biological basis of consciousness have consequently taken their inspiration not so much from specific cerebral regions but from their interconnectivity. A lot of research has focused on the reciprocal loop between the outer layer of the brain, the cortex, and the deeply situated thalamus, an area which acts as a kind of relay hub for the different senses: the thalamocortical loop. Why should this circuit in the brain be so attractive to researchers? There are several reasons.

The first is that the thalamocortical connection does not function in patients in a persistent vegetative state such as a coma, suggesting that it might well have a key role in consciousness (although it is also the case that there is less activity in a range of areas when a patient is in this state). Second, in the early stages of sleep, the cells in the thalamus and cortex are much less active compared to the waking state. Third, researchers have found that direct applications of an arousing chemical (nicotine), which simulates a chemical messenger in the thalamocortical loop, can restore consciousness;[25] conversely, damage to the thalamus results in its loss.[26] Finally, anaesthesia may abolish consciousness because the thalamic cells are no longer driven by feedback returning from the cortex, because the loop circuit is disconnected.[27] But despite all this, it is still unclear why the thalamo-cortico-thalamic circuit isn't simply at most necessary, as opposed to sufficient, for consciousness. Where and what could the special key processes in consciousness be that differentiate this loop from all other connections in the brain?

A more helpful possibility for meeting the requirement of the transition from consciousness to unconsciousness could be to look at the actual state of activity of the cells in question. During sleep, the cortex undergoes periods of short bursts of excitation; these alternate with spans of quieter activity. It is this fluctuation in thalamocortical networks that has been linked to the transition from consciousness to unconsciousness. When the thalamus cells are quiescent, it would mean that the target cells in the cortex revert to some kind of ticking-over, default mode: the characteristic slow brain waves that underlie unconsciousness.[28]

Instead of 'simply' monitoring which brain regions may be active in certain circumstances, a later game plan focuses on more sophisticated mechanisms as a more persuasive and faithful parallel to consciousness: where and when brain regions are interacting, in a kind of cross-talk of large-scale, reverberating oscillations. Some of these widespread oscillations relate to relaxation and sleep and depend on long-range interactions *between* the thalamus and the cortex. Meanwhile, other frequencies seem to be generated *within* specific areas of the thalamus or cortex and these have been linked to higher cognitive functions, including perception.[29] If collective banks of neurons temporarily work in synchrony, this process could have tempting functional possibilities: it turns out that one specific frequency – 40Hz – is a particularly promising candidate for the much-sought-after correlate.[30]

However, more recent research has suggested that such synchrony between neurons might actually be the standard mode for the brain rather than the exception,[31] and there is no reason for viewing any one particular frequency as the sufficient, mysteriously privileged condition for consciousness to appear. Furthermore, unconsciousness can be accompanied by *increased* synchrony in the sophisticated, frontal part of the brain.[32] Finally, there are many conscious experiences that have *not* been associated with synchronous brain activity, while highly synchronized activity, such as that which occurs during a seizure, is in fact linked to a loss of consciousness.[33]

What all these different studies show us is that the strong anatomical reciprocity between the thalamus and the cortex provides, at best, a necessary rather than a sufficient basis for consciousness and, at worst, raises more questions than it answers. A recurring problem in finding the elusive yet significant NCC is that many scientists tilt towards what they know best – the 'N', the neural – and spend little time on the corresponding 'C', the consciousness bit. As it stands, a neural correlate of consciousness consisting of thalamocortical loops, which incidentally happen to be standard issue in the brains of all mammals, fails to account for any of the phenomenological features of consciousness such as the effects of mood-modifying drugs and species differences – let alone more fine-grained variations in individual human consciousness and the fact that we must find a unique neuronal equivalent of our experiences every progressive moment we are awake. Many studies highlight the necessity of various brain regions but do not suggest *why* they should be special in contributing to, or being responsible for, consciousness, above and beyond the fact that they are part of a complex interactivity between brain regions.

An alternative line of attack could be to focus not so much on space and spatial features of the brain but on time as a way forward:[34] it seems that consciousness can be possible only when neural activity is sustained for a prolonged period, some hundreds of milliseconds. The crucial cut-off for distinguishing between 'seen' and 'unseen' events is apparently between 270 and 500msec (thousandths of a second):[35] the innovative neuroscientist Benjamin Libet suggested that such a period was necessary for consciousness of an event somehow to build up, even though the earliest responses might already have been recorded in the brain by as early as 25msec.[36]

An alternative interpretation could be that the crucial factor might not be time itself but the *end* to which a long time-frame is the *means*: this end would be the creation of an appropriate window to allow 're-entry', a kind of continuing reverberation of input and output between specific brain regions.[37] An important tenet of this theory is that no one should confuse 're-entry' with simple feedback, whereby a certain existing state is modified as a result of an initial effect. Monkeys apparently respond that they have 'seen' an experimental figure only when their brain waves indicate a characteristic pattern which reflects this kind of neuronal boot-strapping ('re-entry'),[38] something which also occurs in human subjects.[39] Yet while this suggestion seems plausible, it tells us little about why 're-entry' – this building up of iterations between two brain regions – should be of such primary importance.

So let's change tack again. In the hunt for a convincing link between consciousness and a particular brain property, we can also travel in the opposite direction by scaling down anatomically and focusing not on the connections between brain regions but on their smallest components: individual brain cells. With this approach we can monitor single neurons in humans undergoing brain surgery. Gory though this sounds, there are no pain sensors within the brain, so it has been quite feasible since the mid-twentieth century for a neurosurgical patient to be fully conscious while their brain is stimulated, as we have seen already with the astonishing work of Canadian surgeon Wilder Penfield.[40] Penfield, who in his time treated some five hundred patients, was able to stimulate the exposed surface on the side of the brain (temporal lobe) while the patient was undergoing surgery for severe epilepsy. Although, often, the procedure was uneventful, sometimes, amazingly, patients gave spontaneous reports afterwards of having experienced vivid yet dream-like 'memories'. The same 'memory' could be evoked when other nearby sites were stimulated, while stimulation at different times at the same site could result in different subjective experiences. These revelations suggest that Penfield was tapping into diverse, albeit overlapping networks of neurons in which one neuronal site might be part of more than one network and, conversely, different sites might be part of the same network.

Despite this clear indication that neurons do not work in isolation, it is only recently that a more refined technique on awake human patients not under general anaesthesia has been developed to record the activity of single brains cells, with utterly astonishing results. In one experiment, which at first might seem a little weird, a single brain cell in a neurosurgical patient was activated by as many as seven different pictures of Jennifer Aniston – but this same cell was unresponsive to as many as eighty other images, including those of other comparable and contemporary film stars such as Julia Roberts, or even pictures of Jennifer Aniston with Brad Pitt. There were certain brain cells that responded by specifically increasing their activity only when the subject was shown pictures of a particular celebrity. In another example, the researchers found that about a third of the cells they tested (44 out of 137) were highly selective in just the same way: in this case, the photos used were of Halle Berry and the particular brain cell was henceforth dubbed the 'Halle Berry neuron'.[41] What impressed the scientists was the consistency of responses even when the subject was shown highly distinct and varied photographs of the same person or object, a phenomenon known as 'invariance'. Such studies suggest that there is cognitive processing at the level of single neurons. But could a single brain cell really act as some kind of independent entity and arbiter – in other words, as a neuronal Supreme Controller? Let's unpack this bizarre scenario a little more . . .

In the mid-twentieth century the easiest way of thinking about how the brain was organized was to picture it as a kind of hierarchy, similar to a chain of command, with the boss at the top of a pyramid-like structure. This concept fitted well with scientific findings of the 1960s, when two physiologists, David Hubel and Torsten Wiesel, made a breakthrough that was to win them the Nobel Prize twenty years later.[42] Hubel and Wiesel were working on the visual system and monitoring the activity of single brain cells in the various brain regions which processed inputs from the retina, and their activity on into the depths of the brain. Their remarkable finding was that, as they probed deeper into the brain, further away from the initial processing of the retina, so the cells seemed to become fussier in terms, literally, of what turned them on.

The sight of any old blob would excite a neuron low down in the pecking order, but higher up in the visual-system chain of command it might have to be a line, and higher than that only a line in a certain orientation, and higher still a line in a certain orientation but this time moving only in a specific direction. So it seemed that there was a hierarchy of sophistication in processing visual information: as the input from the retina was processed by subsequent stages deeper into the brain, so the cells in question became less generic and more specialized. It was certainly an astonishing discovery that a single brain cell could have such an individual signature: but it gave rise to some strange extrapolations that went beyond the reasonable idea of a hierarchy of ever more defined physical features in the brain to the far-fetched idea of a hierarchy of cognition and consciousness.

The further up in the hierarchy of the brain you advance, the fussier the brain cell becomes, eventually responding only to very sophisticated and selective images, such as a face, even a particular face. Back then, scientists referred to a hypothetical 'grandmother cell' which, as its name suggests, would respond only to the sight of your grandmother, as the ultimate stage in the hierarchy, the final level of processing.[43] However, the idea that a single 'grandmother cell' could effectively be a discerning mini-brain in its own right has been largely discredited – if only because we could never have enough brain cells to represent 'all possible concepts and their variations'.[44] Then there's simple logic: if you never had a grandmother, a cell would be redundant and wasted: or if you did have a grandmother but your grandmother cell died, as many neurons do daily, then you wouldn't recognize your grandmother ever again!

A more attractive theory accounting for the apparent ingenuity of these single neurons is that, somehow, the more sophisticated, later-stage processing of vision in the outer layer of the brain (the visual cortex) is able to convert different visual data into a uniform, common format (view-invariance) by storing diverse memories of different angles of the same face or object at different times.[45] However, even such cerebral wizardry cannot be the full story. It turns out that some of the cells in the patients were activated not just by a visual image of a face at all but simply by the printed name of the individual in question.[46] So the 'invariance' observed was based on neuronal connections relating to memory, way beyond the processing of vision alone.

If we revisit the research done in 2005 giving rise to the notion of the Halle Berry neuron, it might be best to think of such a neuron – or another, equally fussy one – neither as being unique nor indeed functioning in isolation. If Halle Berry were in some way represented in the brain by a very large network of cells spanning large swathes of the cortex – as suggested so long ago by the work of Penfield – then the problem of the Supreme Controller would be circumvented and an explanation for the activation of the cell by non-photographic stimuli made possible. However, although such studies give great insight into how the brain might be personalized, and how it adapts to an individual's experience, once again they do not help us to understand consciousness itself. And yet, candidate correlates of consciousness haven't even stopped at the level of single neurons but can be on an even smaller scale – *much* smaller . . .

Some fifteen years ago the mathematician Roger Penrose and the anaesthetist Stuart Hameroff came up with a completely different approach.[47] Their rationale was that, because consciousness has not yet been satisfactorily described in terms of any step-by-step computations (algorithms), some kind of novel, non-algorithmic brain process was needed: a new sort of quantum physics (the science of the very small), in which phenomena do not behave in the way the traditional large-scale theories first developed by Isaac Newton would predict. The suggestion involved neither brain regions, nor their connections, nor even a single brain cell: instead, Penrose and Hameroff's account started with 'microtubules', microscopic rigid yet hollow rods which are *inside* every cell. These microtubules are incessantly changing their structure – forming, dismantling and re-forming: as such, so the idea ran, their changeable configurations would correspond well to the possible holistic states of a system (according to the accepted principles of quantum mechanics). Once the number of neurons recruited was sufficiently large, the laws of an as yet unrecognized new kind of quantum physics would take over and trigger a sudden shift to one particular physical state, dubbed 'quantum coherence', which would correspond, in some way and for some reason, to a moment of consciousness within the brain.[48]

There is a fundamental quibble to be had with the Penrose and Hameroff scheme. The microtubules which would drive the 'new kind of physics' are generic to all cells, but only in this case are they said to operate within neurons as a site for a quantum event linked to consciousness. If so, what would be the constraining, additional special feature within the brain that enables them to behave in such a very special way? More fundamentally still, this theory raises as many questions as it answers:[49] not least among these is the simple reasoning that, just because the brain doesn't exclusively obey algorithmic principles, it doesn't mean that a theory offering a non-algorithmic process must be automatically synonymous with the process of consciousness.[50]

Let's take stock. Although various candidate NCCs may be necessary for consciousness, none has been proven, on its own, to do the job: no special brain area, circuit or bank of cells removed in isolation to a Petri dish would seem not just necessary but also independently sufficient for consciousness to ensue. Perhaps it is inevitable that neuroscience-based approaches to NCCs emphasize the neural at the expense of the conscious. Luckily, we have an alternative: instead of starting with the physical brain and coming up with an exotic or interesting property that is sufficiently novel to attribute to it consciousness-driving powers, we can instead tackle the problem from the other way around. In this approach, we will need to come up with a theoretical 'idea' or 'model' for consciousness and then determine whether it could be convincingly validated by neuroscience.

FROM THEORY TO EXPERIMENT

One of the first theoretical models for describing consciousness was based on the concept of a kind of 'blackboard' in the brain, a 'global workspace'.[51] This theory was initially developed by Staneslas Dehaene and his colleagues in the late 1990s, where functions were hypothesized to be coordinated and processed between multiple inputs, all of which fed into some kind of transitory common neuronal 'forum', or 'stage', from one moment to the next. This scenario is one of a global state that, just for one moment, is so dominant within the brain that its very existence prevents the formation of any other, thus producing a single state of consciousness.

This defining feature of a single dominant state in the brain which then matches up with consciousness has inspired the philosopher Daniel Dennett to suggest what he called the 'multiple drafts model', whereby it will be possible to determine what state will indeed be dominant, that is, feature as the content of that moment of consciousness. He suggests a kind of competition, or survival of the fittest, in which the deciding factor for domination is that of the most significance: that which has the most 'fame in the brain'.[52] Yet whether or not the content of a conscious experience is so strong, or extensive, or significant, as to trump all other contenders, the buck has to stop somewhere in the brain to be finally appreciated. But where and how? As we saw earlier with the Halle Berry neuron, there is no overriding neuronal controller. The global-workspace and multiple-drafts models are of little help in understanding consciousness, other than to introduce an element of democracy where, rather than there being any fixed place or predetermined brain feature that gives rise to consciousness, a kind of unspecified neuronal majority wins the day on a moment-to-moment basis.

Alternatively, instead of viewing consciousness as a process, a verb, we could see it as a noun, an entity. This idea might seem just downright silly. However, ever since French physician and philosopher Julien Offroy de la Mettrie suggested in 1747 that 'the brain secretes thought as the liver secretes bile', certain thinkers have debated whether consciousness might in fact be a 'thing'. This is the position of the Panpsychics, a school of philosophy which asserts that consciousness is indeed an entity, a distinct phenomenon out there, an irreducible property of the universe, with the brain acting as some kind of satellite dish picking up the ethereal signal.[53] So, just as time and space exist independently of the brain but can be appreciated by it, consciousness, likewise, is an independent entity which may or may not be detected as and when brains happen to be around. However, there's no point here in going further into the arguments for or against this idea, simply because, from a neuroscience perspective, it is not very helpful, and there's no obvious way of progressing from it.

More recently, however, an alternative and more abstract concept has been suggested by neuroscientist Giulio Tononi,[54] who has developed a model, the Integrated Information Theory, where 'integrated information' is the reduction in uncertainty about the state of a variable within a system, and consciousness is in proportion to how many other states can be ruled out.[55] However, as with the earlier concept of 're-entry', it is hard to see what the mere notion of reducing possibilities alone adds to our understanding of what consciousness is and how it relates to the physical brain. The integrated information model does have the advantage over the earlier 'quantitative' ones in that it can be simulated more readily and accurately on a computer, but it would still teach us nothing about the conversion of the water of the physical brain into the wine of subjective experience,[56] or even the 'mere' correlation of the two.

We could be focusing, as we've seen, on any number of brain features: widespread, albeit ill-defined, brain 'activity',[57] or the biggest 'impact'[58] of competing processes, or permissive time frames,[59] or iterative 're-entrant' neuronal communication[60] or, as here, integrated information: but if so – and this is the catch – then any feature distinguished, as these all are, by sheer *quantity*, must at some stage be translated into the oh-so-elusive special *quality* of subjective conscious experience. Why and how should crossing a *quantitative* Rubicon give rise to a distinctive *qualitative* inner state?

The brain regions selected as being linked to consciousness function at the cellular level in the same way as those that are less sophisticated and not singled out as such: so it must be the emergent property of the complexity itself that makes the crucial difference. Indeed, some people, such as the technologist and futurist Ray Kurzweil put all their money on complexity, independent of any biology: in 2012, Kurzweil suggested that 'Artificial intelligence will reach human levels by around 2029. Follow that out further to, say, 2045, we will have multiplied the intelligence, the human biological machine intelligence of our civilization, a billion-fold.'[61]

Would Kurzweil envisage consciousness as still being an attractive part of this new world order? If so, it follows that by building machines of ever greater complexity, consciousness will emerge as a spontaneous and inevitable rabbit out of a computational hat.[62] If the premium is on 'complexity' exclusively and unconditionally, then consciousness would not need to be an exclusive property of biological systems but could be made of anything, so long as it was 'complex' enough – as the philosopher John Searle once quipped, even old beer cans. The material itself wouldn't matter, only the interrelations of its component parts.

However, from the perspective of neuroscience, this way of progressing with complex and/or computational systems irrespective of the material from which they are constituted is missing something: the trafficking of the huge variety of capricious and powerful compounds in the nervous system that work in different combinations, in different places, over different windows of time, with highly context-dependent and variable effects. Each neuron is *not* like a beer can but is in fact highly dynamic: the 100 billion highly changeable neurons that make up your brain are far from being fixed components that will plug in and play consistently, independent of the surrounding environment in which they are located. Moreover, accommodating this functional dynamism are incessant anatomical changes in the configuration and shape of each neuron: the ease by which signals coming in to these cells vary greatly from moment to moment according to the availability of diverse, qualitatively discernible 'modulating' compounds swirling around that particular brain cell.[63] This intense, ever-changing kaleidoscope of interacting neurochemicals and neuronal structure is nothing like the rigid circuitry of computational devices.

Moreover, there's a whole body out there, with incessant feedback trafficking to and from the brain. Almost twenty years ago the neurologist Antonio Damasio pointed out the importance of chemical signals that feed back and forth between the brain and the rest of the body, chemicals he referred to as 'somatic markers'.[64] The interplay between the three great control systems of the body – the immune, endocrine and nervous systems – should never be ignored. After all, if they were not interactive, there'd be biological anarchy, and we wouldn't be faced with the well acknowledged albeit baffling phenomenon of the placebo effect. The diverse neurochemistry of the central nervous system, and indeed of the rest of the body, functions in a way whereby biological quality cannot be reduced to an abstracted quantity, to mere computation. Or, if it can, then that 'model' should be argued and justified much more persuasively than it has been, rather than just swallowed wholesale as an article of scientific faith. In general, the more an approach to consciousness is abstract and theoretical, the less it correlates to real brain events, and the more expectation there is that it should have greater predictive power and that, as an explanation, it should therefore go further and, hypothetically at least, tackle what consciousness actually is.[65] We don't want a 'model' that will beg the question, a theory already predicated on an assumption of what consciousness is, any more than we might want a candidate NCC that just happens to be observed in the lab as an unexpected or gimmicky property of the brain. Instead of NCCs in isolation, or abstracted models lacking validation by subjective experience, it is time for a completely different approach.

A WAY FORWARD?

We really have to take on the most basic question of all, of how the 'water' of objective brain events is transformed into the 'wine' of subjective consciousness[66] – the so-called 'hard problem':[67] but then it is indeed hard to see how it can be solved with the approaches used so far, either going from experiments to theory, or vice versa. We need to move away from simplistic studies of perhaps necessary but persistently insufficient NCCs, and from free-floating hypothetical models, towards uncovering a more faithful and detailed relationship between objective physiology and subjective phenomenology. Just how we might achieve a better picture of this relationship, though, is far from obvious.

The problem is rather that the nature of the mental seems utterly distinct – almost another language – from that of the physical processes in the neuroscience toolkit. How can the first-person perspective of consciousness and the third-person perspective of physical brain states be descriptions of the very same overarching process, and one that should therefore be expressed bilingually? Descartes took these considerations to indicate that mind and brain are in fact wholly distinct substances, but that radical division no longer seems tenable. It is clear that the objective and subjective are, at the least, very intimately related, and we need to determine just what that intimate relation might be. The challenge is to understand the relations between the mental properties of being conscious and the physical properties of brains, *where both physiology (the objective perspective) and phenomenology (subjectivity) are given equal weight and constantly cross-referenced against each other.*

If we can establish accurate neural correlates, and it's important to recognize that there will indeed be more than one, of moments of consciousness, then we can better understand how the phenomenological corresponds to the physiological – even though the *causal* connection still eludes us. It would be a real advance, surely, to show that certain conscious moments can be linked to respectively varied and specifically objective scenarios in the brain and body, be it some kind of neural activity or a release of chemicals. We need to establish a new approach that gives equal weight to objective and subjective events, a common third 'language' whereby the objective and the subjective can be expressed with equal ease and reciprocity. The different approach here in this book will be to start with real-life phenomenology as a clue as to what this third language, or common coinage, could be. Rather than immediately burrowing into the brain and peering at the brain cells and the chemicals, hoping that we'll be able to come up with The Answer, we should first figure out what we would require of any account of everyday consciousness.

So, before we do anything else, we need a shopping list from the diversity of real life, not one compiled in a contrived laboratory situation in which crucial elements may have been standardized, reduced to a minimum, or omitted altogether. Rather than forcing this kind of simplified phenomenology on the brain, as happens in the lab, we should see how it might match up, in physical terms, to the wide variety of experiences that is on offer from everyday life. We can then look for clues as to how the physical brain could cater for these various events which characterize different subjective experiences and see if any general principles emerge.

What are these different features and puzzles of the brain that we have to account for in order to understand consciousness? Here is a list that is probably not exhaustive but captures some of the more essential issues.

First, why does an alarm clock wake you up? This might seem like a basic question, but it is one of the undeniable fundamental features of consciousness. Second, what is the difference between non-human and human consciousness and, by the same token, between self-consciousness and infant-like consciousness? Third, why do the inputs conveying sight and sound enable each individual to have very different subjective experiences of vision and hearing? Fourth, how does the environment impact on our consciousness? Fifth, how do mental conditions like schizophrenia, depression and Alzheimer's, and substances such as drugs and alcohol, change our consciousness? Sixth, how does the subjective experience of a dream differ from that of wakefulness? And, finally, why is it that our consciousness of time passing varies so much?

Only once we have examined these issues can the second step be a neuroscientific one: to determine what we are going to measure in the brain. Instead of selecting a single feature from the outset, whether it be thalamocortical re-entry or quantum coherent microtubules, now we can use the shopping list of everyday experiences to guide us. By considering each item on the above list in turn, we can then explore how the brain can deliver what is needed to account for such phenomenology. As it turns out, each item conveniently corresponds to different stages of a 'typical' day, so 'you' will be our guide. And as we track 'your' day we will constantly switch back and forth between the physical and the phenomenological, the objective and the subjective.

Outside, the sky is brightening. There is no time to lose. Already, the alarm is bleeping . . .

2
Waking Up

The shrill bleep of the alarm is piercing your skull. Little by little, the foggy comfort insulating you from the outside world is evaporating as the intrusion becomes ever more insistent. Stretching and groping with a clumsy hand for the offending object, you are finally able to restore silence. But the fiendish device has already done its job: you are awake. Yet, for the moment, you are still far from being 'all there'. Your eyes still closed, only gradually do you feel that you are surfacing into consciousness . . .

SLEEP

Perhaps the realization that your consciousness could, moment by moment, be slowly growing shouldn't be that surprising: we all know what it feels like to wake from a particularly deep sleep and biologists have long known that the unconsciousness of sleep comes in degrees varying in 'depth'. In fact, over the previous night, you'll have gone through some five cycles of unconsciousness, progressing repeatedly up and down from light to deep levels. Just as you drift off, for the first five or ten minutes, you'll still be relatively alert – in a transition stage between wakefulness and slumber. If someone had tried to rouse you, you might even have said that you weren't, in any case, truly asleep. This is Stage One, the start of the cycle, and the beginning of your descent into unconsciousness. Occasionally during this period you'll experience strange and extremely vivid sensations, as though you are falling, or hearing someone call your name. Sometimes, your body acts outs what your wife has teased you is a 'kick-start into dreamland'. This reflex is formally known as a 'myoclonic jerk': your legs involuntarily jolt, seemingly for no reason at all. Throughout Stage One, if you had electrodes on your scalp, the electroencephalogram (EEG) would be recording a characteristic pattern of small and fast brainwaves (theta waves).[1]

As you begin to relax more, over the next twenty minutes your brain starts to generate further complex electrical wave signatures (eight to fifteen cycles per second; a 'sleep spindle'), whereby each successive wave waxes and wanes in amplitude and where, having peaked, decreases once again: this is Stage Two. Your body temperature is dropping by now and your heart rate is beginning to slow down. It's at this point that you are transitioning from light into deep sleep: Stage Three. But now your EEG brain-wave profile slows down even more, into an unhurried pattern of two to four cycles per second. Once these 'delta waves' reduce further, to half to two cycles per second, you have entered the deepest sleep stage of all, Stage Four.

After some thirty minutes, your brain recycles itself to Stage Three and then back on up to Stage Two. Why? Surely it would be easiest for unconsciousness just to remain as a simple steady state. One possibility is that by cycling the depth of sleep, the brain is protected from being at its most insensitive for a long period of time: a deep, comatose state could be counter-productive to maintaining internal body functions over a prolonged interval, as well as rendering you, or any animal, less reactive to possible external perils such as predators for dangerously long intervals. In any event, an interesting clue to the rationale for cycling depth of sleep, is that the cycles themselves change in length as time unfolds: perhaps the brain has shifting needs as the night wears on . . . By midway through the night, Stage Four has disappeared altogether, while the first cycle of a further stage – Stage Five – initially only lasting about ten minutes, has become progressively dominant, now extending for almost up to an hour.[2]

Stage Five is one of the most well-known features of sleep and is also called Rapid Eye Movement (REM) sleep, because your eyes dart back and forth beneath your closed eyelids. Meanwhile, your respiration rate has increased and your EEG has revealed a profile of fast, irregular waves, indicative of enhanced mental processing and comparable to when you are awake: dreaming (which we shall explore in its own right in more detail later) takes place during this time, though not exclusively so. Despite all this hectic brain activity during your dreams, your muscles at this point eventually become more relaxed, and an eventual paralysis sets in: hence, REM sleep is also known as Paradoxical Sleep, since you may be having some kind of an inner, conscious experience but at the same time you are immobilized. Just think of the experience that occurs commonly in nightmares of trying to run away from some life-threatening danger yet weirdly feeling rooted to the spot. After a period of REM, your body usually returns to the lighter, Stage-One sleep – then the cycle goes through the stages approximately four or five times throughout the night.[3] How might all this cycling be controlled?

Brain scientists have known for a long time that underlying the five sleep stages, and indeed wakefulness itself, are surges within the brain of a range of specific chemical messengers, transmitters. Transmitters act as intermediaries between one cell and the next by diffusing across the intervening narrow gap (synapse): they then cause the target cell to become more or less active, i. e. 'excited' or 'inhibited'. In neuro-speak, inhibition is simply a decrease in likelihood that a neuron will be able to generate an action potential (an electrical blip); excitation is the reverse – an increase in likelihood. This vital electrical blip lasts for about one thousandth of a second (1msec), and is the universal, all-or-none indication that a brain cell is active and signalling to the next cell along. A neuron that is excited will be generating volleys of action potentials at high speed, while one that is inhibited could well be completely silent.

The transmitters that are at work in sleep, waking, and dreaming (dopamine, noradrenaline, histamine and serotonin) are close relatives when it comes to their molecular structure, along with a fourth (acetylcholine), a slightly more distant cousin: they are probably the best-known and most well-documented transmitters in the brain.[4] But the really interesting issue is the characteristic way they are distributed and located: they could be doing more than simply their classic job of operating across a single synapse in rigid circuits in the brain. Instead, they are organized like spraying fountains: large networks of brain cells, each containing its own respective chemical sibling, cluster near each other within the most primitive hub of the brain (brain stem), just above the spinal cord. From here, the hub cells are able to send distant signals to the 'higher' cerebral regions with long and diffuse connections: the powerful transmitters are unleashed upwards, outwards on to large swathes of sophisticated areas of the brain, and in particular the cortex. It turns out that each member of this chemical family plays its own key role in sleep and wakefulness. Noradrenaline and its chemical parent dopamine, along with their siblings serotonin and histamine, are all most abundant in wakefulness, decline during normal sleep and are virtually absent in REM:[5] meanwhile, acetylcholine will still be surging during dreaming.[6] So what do these diverse transmitters actually do?

It turns out that these molecules, so familiar as transmitters, can lead a double life and operate in a very different alternative role all together – as modulators.[7] A modulator doesn't just take a single message across the synapse and cause immediate inhibition or excitation: instead, it will influence how a brain cell will respond to some input within a time frame in the future without having an effect itself right there and then. One way to think about this process is to imagine a situation in an office, say, where there's the rumour of a pay rise. The rumour in itself won't change outward behaviour – for example, no one will pick up a silent phone; however, when a standardized input occurs, say the phone rings, the employees may pick up the call with greater alacrity than otherwise. A modulator would act a bit like the rumour: in itself, it is ineffective but it amplifies a subsequent event.[8]

The concept, and indeed the reality, of neuromodulation, shows that it is misleading to refer to transmitters, as some neuroscientists still do, as definitively either excitatory or inhibitory, as though the function were predetermined and locked into their structure. It all depends on the timing and the particular micro-landscape within the brain in which they are working. Within a certain period of time, the effects on a neuron of an incoming stimulus (another transmitter) will be different in the presence of a modulator, compared to a time when the modulator is absent, or, indeed, if the second transmitter didn't appear at all. The timing – the contingency of modulator and second input – is all-important. So, since the effect of an otherwise consistent stimulus will now be converted into a variable one, the enormous value of modulation is to supply a *time frame* to brain operations that would never be possible with simple, one-off transmission.

Meanwhile, while you have been sleeping – and, indeed, now that you are half awake – levels of these all-pervasive key modulators have been rising and falling at different times, underpinning the different sleep stages by predisposing large populations of your brain cells to be more active or more silent. It is therefore highly likely that in so doing these chemical fountains are making an important contribution to the states of consciousness and unconsciousness, and the transition between the two. But consciousness seems not, after all, to be an on–off switch. We've just seen that, instead, the passage into and out of sleep is a gradual process: so it could be that the modulating fountains act not so much to abolish or trigger consciousness completely, in the manner of an on–off light switch, but, rather, they will act like a kind of dimmer control . . .

ANAESTHESIA

Since sleep is a gradual process, it may not be so surprising that another familiar form of unconsciousness can also vary in depth: anaesthesia. Henry Hickman, the 'father of anaesthesia', first reported the consciousness-robbing effects of carbon dioxide in the 1820s; however, it wasn't until 1937 that an American physician, Arthur Ernest Guedel, described the four stages of anaesthesia which are still used as a framework of reference today.[9] Back in the mid-twentieth century, inhalational anaesthetics were far less efficient than they are now and therefore induction of anaesthesia was relatively slow: this shortcoming nonetheless helped uncover what Guedel demonstrated as identifiable stages in the gradual process of losing consciousness. These days, with rapid intravenous induction of a state of anaesthesia, the steps are not so conspicuous, but they still unfold, albeit more rapidly.

The first stage you experience is Analgesia (from the Greek for 'absence of pain'): this state can be confirmed by the loss of a withdrawal response – say of your arm to a pinprick on the skin. During this stage, any previous pain you felt will be alleviated; you probably won't even notice that you no longer feel pain, although you might carry on talking . . . Next, as you lose consciousness, you enter the second stage of anaesthesia, in which your body will nonetheless exhibit signs of Excitement and Delirium. Your pupils become dilated and your respiration and heart rate irregular; you may also experience uncontrolled, involuntary movements and even, on very rare occasions, vomiting. As you enter the third stage, Surgical Anaesthesia, your muscles relax and your breathing slows right down. Your eyes, which initially were rolling, become fixed, you lose your corneal reflex (the blink reflex to touch of the eye), as well as the response of pupil contraction to the stimulus of light. Your breathing becomes shallower still. At last, you are deeply unconscious and finally ready for surgery.[10]

So anaesthesia, like sleep, is a gradual process. The slow, step-by-step effects of loss of consciousness before surgery can nowadays still be revealed with an analysis of a patient's EEG using a method introduced some twenty years ago. The procedure attempts to give a read-out number – a Bispectral Index (BIS) – as a measure of the level of consciousness (or, rather, unconsciousness) during anaesthesia in surgery. The idea behind the development of BIS was to avoid the nightmare scenarios of either too much anaesthesia causing death or too little: the patient would still be semi-awake yet alarmingly unable to report the fact, thanks to muscle relaxants causing paralysis and therefore preventing speech.[11] The whole point of mentioning BIS here is that, however imperfectly, unconsciousness can in some cases apparently be quantified beyond simple all-or-none states.

However, the problem with BIS is that it does not give the same sensitive read-out for *all* anaesthetics. In order to really understand the brain mechanisms involved, therefore, we need to square up to this apparent paradox: on the one hand, different anaesthetics must be working by different neuronal processes, and therefore on different parts of the brain, but on the other hand, they all nonetheless lead you to the same endpoint – a uniform loss of consciousness. Among anaesthetists, there are two competing theories that are attempting to resolve this riddle. One is that there is a final common pathway underlying loss of consciousness; the other that there are just many different brain states with crudely similar outward features. The main issue bedevilling a real understanding of which of these two scenarios is the more likely is that the techniques for comparing the anaesthetics in question differ so it is hard to compare results in an even-handed way.[12] Yet, although the discovery of any common mechanism of unconsciousness is an urgent priority, the most immediate point here is the undeniable fact that unconsciousness itself is variable in depth.

CONSCIOUSNESS AS VARIABLE

If unconsciousness comes in degrees, both in sleep and anaesthesia, *then might consciousness itself also come in degrees, and be continuously variable*? If consciousness really is changeable, then various riddles could become easier to answer. For example: is a foetus conscious?

By just the end of the fourth week of gestation, the human embryo already has a brain in three distinct parts (the forebrain, midbrain and hindbrain), which begins functioning within the following week.[13] So is this tiny being conscious? And if it is, then when, and how? Just assume for a moment that the foetus is *not* conscious. But then, when would things change? Perhaps when the baby eventually squeezes down the birth canal. If so, you would face a pretty tough prospect if you happened to be born by Caesarean section: you would never be conscious throughout your entire life. So instead, perhaps, the crucial, deciding factor is arrival at full term of a pregnancy, at forty weeks. Yet it is a hard scenario to imagine the new parents of a premature baby saying, 'Look, nine months now. The baby wasn't conscious yesterday but will become conscious today. Now, at last, it's worth visiting the hospital.' Clearly, these are crazy scenarios, not only in terms of common sense and plausibility but because of the hard fact that the brain is indifferent as to whether it receives oxygen through its mother's umbilical cord or its own nose. So the pressing question then is: when does a foetus actually become conscious? There is, after all, no clear line in development, no neuro-Rubicon that is crossed as the brain grows in the womb – no single event or change in brain physiology, and certainly not at birth, to suddenly trigger consciousness.

A more realistic, and scientific, approach would be therefore to reject altogether the suggestion that a foetus has never been conscious. With it, we can also jettison the uncomfortable notion that consciousness is some kind of qualitative magic bullet. Let's instead go back to the image of the dimmer switch and think of consciousness as a consequence of the *quantitative* increase in brain volume and density during development. In other words, consciousness could grow as biological brains grow, both in the womb and in evolution.[14]

If consciousness is indeed continuously variable, it would mean that you, as an adult human being, could be more or less conscious from one moment to the next. We talk about 'raising' or 'deepening' our consciousness: it doesn't matter whether we go up or down but, effectively, we'll have been tacitly assuming degrees – amounts – of consciousness. Why is that so helpful? Well, we can now do something that scientists find reassuring: we can at last measure *something*, whatever it may turn out to be. We can now convert a still admittedly unidentified brain phenomenon that was seemingly qualitative into something quantitative which, if we are lucky, we might eventually be able to measure in the brain. What if we could find some kind of as yet unidentified process that comes in degrees, is continuously variable? If we could do that, then we might have a more constructive approach to discovering a worthwhile and meaningful correlate. So now, finally, we've got something, albeit something still very vague, that we can put on the shopping list and ask the brain to deliver. Where do we go from here in identifying the 'something'?

The obvious place for a brain scientist to start would be with the macro-scale brain regions such as the frontal cortex, thalamus or hippocampus: after all, they are the easiest to see with the naked eye and their shifting patterns of activity are now familiar as the beautiful coloured images seen in brain scans. Accordingly, the diverse areas that are active when consciousness occurs[15] have been painstakingly documented and listed: perhaps not surprisingly, they range all over the brain.[16] An immediate stumbling block here is that the actions of anaesthetics upon inhibiting a certain specific area do not necessarily result in unconsciousness.[17] So the inevitable deduction is that it is a key *complex* of brain regions that need to be deactivated collectively to guarantee finally a loss of consciousness. Clearly, unconsciousness – and, by extension, consciousness – will depend more on the *relationship* between various brain regions.[18] In support of this idea, research shows that during deep sleep the crucial change is when the connectivity between brain regions breaks down so that communication across the brain is less effective.[19] But it is one thing to claim that connectivity between brain regions is all-important – which it clearly is – and quite another to demonstrate, as we saw in Chapter 1, what the key, minimum-kit circuitry or process may be, and where it is. Moreover, the criterion of all-or-none connection versus disconnection does not tick the box for the very first item now on our shopping list: that the unconsciousness of both sleep and anaesthesia is graded, and that the crucial processes operating during these states will also probably have to be correspondingly graded as well.

How can we square the circle of finding a graded, variable process? The only answer is that the basis of unconsciousness would have to lie not in some kind of brain switch either within or between key brain regions such as the cortex or thalamus but in a process that has not yet been really defined in terms of textbook neuroanatomy at all, and which is not describable in terms of an all-or-none electrical blip.

When I first started studying neuroscience, it was common practice to try to explain brain processes by drawing schemes in which brain regions were depicted as tidy boxes with arrows coming and going back and forth between them, with '+' or '–' beside each, denoting a simple net excitation or inhibition. However, we've just seen how much of neuronal communication would be 'modulatory': any eventual inhibition or excitation between or within groups of cells will be dependent from one moment to the next on the *context* of the ongoing status of the cell. This means we need to search for some additional kind of brain mechanism that can operate at a level beyond that of single neurons, which, as we've seen already, do not function as autonomous units. Moreover, whatever it might be, this novel brain process needs, above all, to be something that is analogue: continuously variable from one moment to the next.

NEURONAL ASSEMBLIES

Way back in 1949, the trail-blazing Canadian psychologist Donald Hebb introduced the completely revolutionary idea that neurons could adapt to previous events – in a way, learn. Hebb showed that neurons in close proximity to each other tend to be synchronous, meaning they are active all at once, in a cohesive way: when they are in this state they constitute a unified functional network the operations of which could actually endure well beyond the initial trigger.[20] Hebb went on to suggest that if they were collectively active in this way, then these neuronal networks would have the potential for inducing much longer-term changes at their synapses, which in turn would lead to an enduring, enhanced communication between the brain cells within the network: Hebbian synapses. Why was this scenario such a breakthrough?

The visionary set-up explained for the first time how the brain could adapt to its inputs, and hence its environment – a phenomenon that came to be known as plasticity (from the Greek *plastikos*, to be moulded).[21] Plasticity is now widely acknowledged to be a basic feature of the brain, although different species vary in the extent to which they are liberated by it from instinct. In simpler animals, where behaviour is much more closely linked to genes (say, goldfish), and brains have fewer neuronal connections in the first place, the environment will have less impact than in others where individual experiences will literally be able to leave their mark on the brain. Within the animal kingdom, we humans have the superlative talent for adaptation, which is why we occupy more ecological niches than any other species on the planet – we can live and thrive in more different places around the globe, from the jungle to the Arctic. Arguably, this plasticity of our brains also means that we have the greatest potential, of all species, to become truly unique individuals, by having individual experiences – all thanks to the adaptable Hebbian synapse.

Hebb's theoretical idea was finally proven empirically a few decades later when neuroscientists were able to demonstrate that there was indeed a cellular mechanism of long-lasting, relatively slow adaptation that was, above all, highly localized to individual synapses,[22] and which has had great impact in helping with many explanations of different phenomena in neuroscience and psychology, especially learning and memory. Yet, while this paradigm has been successfully applied and investigated in myriad publications over the last few decades, we still have a problem: either we can explore the brain as Hebb did, at the level of cells, synapses and transmitters (an approach referred to as 'bottom-up') or we can focus on final brain functions and macro-level brain regions (a strategy conversely dubbed 'top-down'). But how do we get from one level to the other?

There still needs to be some kind of way that these small, localized networks of neurons can then in turn impact on the interplay between defined brain regions which gives rise to cognitive processes such as memory, and indeed accounts for the individual we each become. In other words, there needs to be a bridge between the macro-scale (top-down) and the micro-scale (bottom-up). Neighbouring neurons, if and when they are simultaneously active, will thereby enhance their connectivity. But Hebb wondered whether we could go further still. Perhaps this ensuing local activation could, on a much grander scale, ultimately lead to a cohesion in more global activity and function across many more neurons.[23]

Such hypothetical, much larger scale neuronal coalitions were, for many years, impossible to detect or visualize in reality. They would be too extensive to monitor with conventional electrophysiology, which records only a few neurons at a time, and since they had never shown up in typical brain scans, they would be, presumably, not very long-lasting. Remember that brain imaging has a time resolution a thousand or so times slower than the speed at which neuronal communication occurs: just as with those old Victorian photographs, mentioned in Chapter 1, with their slow exposure time limiting their content to static buildings and objects, so all that it has been possible to see with conventional fMRI is ongoing activity over several seconds. Such activity might be helpful in diagnosing ongoing brain malfunction, or in revealing what happens if you slow behaviour down to comply with the timescale by asking a subject to engage in a repetitive, continuous task. But, unlike the localized, 'hard-wired' micro-level synapses or macro-scale brain regions, if they exist at all these hypothetical, mid-level (meso-scale) coalitions would be too temporary to register over the time needed with traditional imaging techniques.

So how could scientists ever know if Hebb's more speculative vision was correct? The challenge was to find a way to link top-down and bottom-up processing: but all that neuroscience had in its toolkit were standard anatomical and imaging techniques for the former, and just a few electrodes for looking at single cells for the latter. Then, in the 1990s, a new technology – voltage-sensitive dye imaging (VSDI) – was developed by ingenious scientists such as Amiram Grinvald at the Weizmann Institute in Israel.[24] With this technology, it suddenly became possible to discern previously undetected phenomena that would have remained hidden with conventional, non-invasive brain imaging.[25] As its name suggests, VSDI gives a read-out of the voltage across the cell membrane, and hence the ongoing activity of neurons: because the dye embeds in the actual membrane, it means that the read-out is direct and therefore on a sub-second, effectively instantaneous, time scale. Using this technique, we can now see for the first time that between the cellular and synaptic levels of operations and anatomically discernible brain regions there is indeed a busy mid-level of brain processing whereby large-scale coalitions of neurons *do* work as a cohesive unit over very fast time scales commensurate with real-time events in the brain.

The figure below is from my own research group in Oxford, where we are working with slices of rat brain using VSDI. Following a brief electrical stimulus, the resultant activity can then be colour-coded: red is highly active, purple less so. Meanwhile, the scale of several millimetres over which this all occurs is quite large compared to a single cell, yet quite small compared to an anatomically defined brain region: a true meso-scale level. Notice in particular the very fast time resolution: within 8 milliseconds, the collective activity is starting to peak and then, eventually everything is almost over, in this case, by about 20 milliseconds. This event could never have been detected with conventional brain scans.

Because these large-scale but highly transient coalitions of neurons are still relatively unfamiliar phenomena and there is as yet no clear

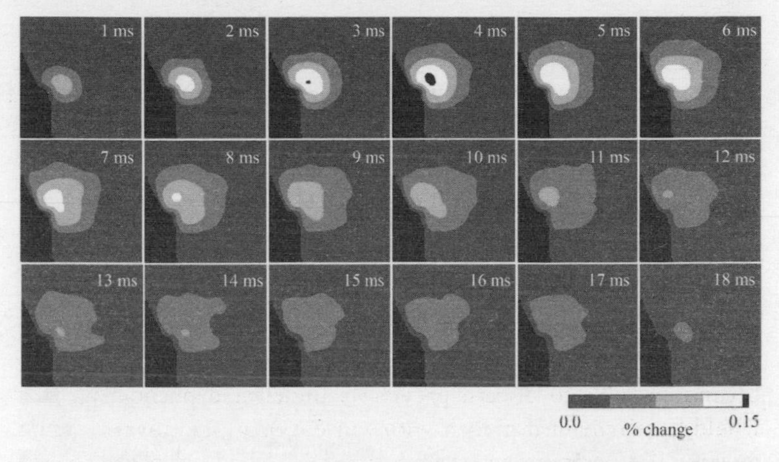

Fig. 1: Visualization of an 'assembly'. Sequence of images one thousandth of a second apart, showing widespread activation, detectable with voltage-sensitive dyes, in a rat-brain slice following an initial pulse of stimulation lasting a tenth of a microsecond. The highest activity is at the centre, gradually decaying to the outer limits – a little like ripples in a puddle from a stone throw. (Badin & Greenfield, unpublished.) (To see this scan in colour, see Plate 1.)

consensus on a rigorous definition, they are referred to by various names. In our group, we call them 'neuronal assemblies', and we define them as: variable, highly transient (sub-second), macro-scale groups of brain cells (for example, about 10 million or more) that are not confined to, or defined by, anatomical brain regions or systems.[26]

As we journey through the waking, working day, I will argue that the only way in which the physical brain could accommodate the ebb and flow of a constantly changing conscious state would be at this intermediate level between macro-scale brain regions and micro-level individual neurons: a mid-level of organization of transient, collective activity of brain cells that expand or diminish from one moment to the next to accommodate varying depths of consciousness. So the idea is that, if consciousness is continuously variable, then it could be linked to the physical phenomena in the brain that we shall term 'assemblies'.

Although imaging in brain slices can reveal the neuronal nuts and bolts of assemblies, it is only looking at the intact brain of a living animal (the technical term is *in vivo*) that will inevitably give more insights into how they work and what they do. In these types of experiments, the initial trigger can be more natural – say, a flash of light,[27] or simply touching the rat's whisker (Figure 2).

In both brain slices and anaesthetized brains, however, the assembly that results is a bit like the ripples in a puddle evoked by throwing a stone. Accordingly, just as a stone is much smaller than the extent of the ripples generated, so, in response to even a brief stimulus of light, activity in an assembly will temporarily spread to a much larger

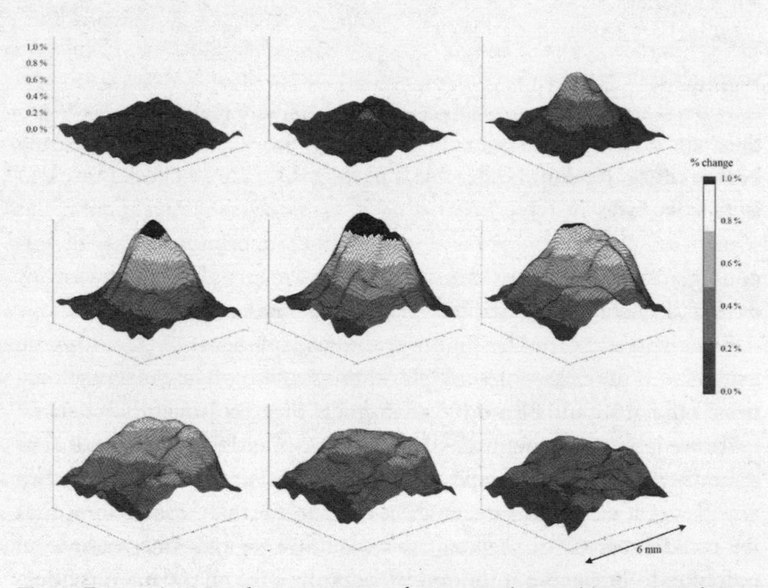

Fig. 2: Frames in 3-D taken every 5ms of a neuronal assembly generated in the intact sensory cortex of the anaesthetized rat, triggered by a whisker deflection. Note that the diameter of the assembly measures some 6–7mm above the background noise.[28] In this case, once again, the time window (40ms) would have been far too fast for conventional brain imaging, and the detailed spatial patterns completely unobtainable with conventional electrophysiological recording. (To see this scan in colour, see Plate 2.)

area,[29] the extent of which is much greater than the arrangement of hard-wired, cookie-cutter circuits of cells in the cortex.[30]

However, when you use optical imaging to look at the brain in the anaesthetized animal, you can also see something else that you wouldn't see normally in brain slices. You'll see that the brain is continuously active even without any obvious stimulation driving it. Large-scale oscillations in activity will be pulsating across the brain: in neuro-speak terms, we know that they can occur due to the different properties of the different cells, such that neurons act as individual oscillators that couple together with different delays.[31] Given the right constellation of properties in a group of neurons, these oscillations could continue indefinitely, providing a background to whatever else occurs as a result of a one-off stimulation or event.[32]

Clearly, the brain is not an uncompromisingly organized structure with distinctive lattices of hard-wired connections like a computer, with specific nodes and binary on/off functional states. While it's undeniable that such connections exist at a localized level in the brain, they are very far off from being the whole story. In other words, the brain is not a rigid and inflexible building; instead, we should think of it as more of a heaving ocean that is sometimes relatively calm but sometimes choppy, churning, even stormy. Superimposed on all this neuronal turbulence are one-off triggers – internal brain drivers or external sensory stimulation that will evoke a one-off assembly.[33]

But could a neuronal assembly fit the requirements of the shopping list we need in order to describe an appropriate correlate of consciousness? If so, it should be modified by agents that abolish consciousness: for example, anaesthetics. We saw earlier that a basic paradox of anaesthetics is that no single process has been identified for how they works: yet it might be here, at the meso-scale level of assemblies, that we could eventually see what otherwise diverse anaesthetics have in common to bring about the single common outcome of unconsciousness. After all, these powerful drugs are available in all shapes and sizes and, in their actual structure, have nothing in common that could define them as a group as against other psychoactive agents, even ones that perform a complementary clinical role such as painkillers (analgesics). In both cases, at the traditional bottom-up level of single cells, anaesthetics and analgesics can have similar effects: a fall in activity, namely inhibition. So where does the all-important functional difference now creep in between drugs that take away pain and those that take away consciousness?

In one experiment, we've investigated whether chemically diverse anaesthetics, compared to painkillers, could have differential effects on the dynamics of assemblies. As it turns out, the painkillers had no impact on assemblies in brain slices but, by startling contrast, the two anaesthetics *did* have an effect: although they are each very different from the other as drugs, both had the same net effect in modifying assemblies by prolonging their duration.[34] So for anaesthetics, irrespective of their different actions at the micro-level within the brain, it is their similar actions at the meso-scale of assemblies that counts.

As we've already established, there's no one-to-one relationship between an isolated area of the brain and consciousness, yet what this finding was telling us is that assemblies are in some way linked to an eventual conscious or unconscious state such that they are specifically impacted by consciousness-blocking agents – but not those that merely modify the content of consciousness, such as analgesics, where you are still awake but no longer in pain. Only anaesthetics that are defined by their ability to take away consciousness prolonged the duration of the assemblies in each case. Clinicians were already aware that prolonged seizures can block consciousness:[35] so here, prolonged assembly activation might also potentially contribute to an eventual similar result. However, brain slices, as we used here, cannot gradually become unconscious, as they were never conscious in the first place. It seems a tough call, but we really, ideally, need a situation in which, with all the convenience of a slice preparation, we could somehow mimic such a transition between unconsciousness and consciousness.

One possibility is to exploit the counterintuitive but long-known fact that when the brain is exposed to extremely high atmospheric pressures (during experiments), amazingly, the consciousness-robbing effects of an ongoing anaesthesia are reversed: the subjects in question, usually tadpoles or mice, wake up.[36] Would a similar change in the immediate environment that restores consciousness itself in an intact organism have any effect on the nuts and bolts of assemblies in brain slices?

Under extreme high pressure, when anaesthesia would be reversed

(32 atmospheres), to our astonishment, the assembly in the brain slice was now considerably larger (see Figure 3), which indicated collective activity of a much larger population of neurons.[37] These results demonstrate that high-pressure conditions *do* have a substantial effect on a particular brain phenomenon (an assembly) that might in turn be a good candidate for a neuronal correlate of consciousness: experimental conditions that dramatically modify the state of consciousness also dramatically modify assemblies in a specific brain area.[38, 39] The next stage is to test the effects of anaesthetics in a more natural set-up.

Accordingly, we've investigated the effects of anaesthesia in a living rat, where what we would be recording would be the overall effect, however indirect, on the brain – indeed, on the whole animal. Because of the invasive nature of optical imaging, the animal, obviously, had to be surgically anaesthetized throughout the experiment. However, thanks to the graded nature of anaesthetic effects, we have

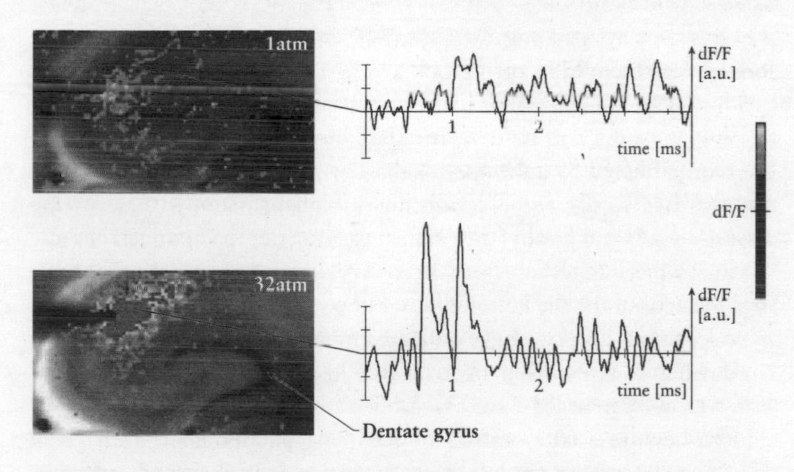

Fig. 3: Fluorescent image of assemblies generated in the hippocampus of a rat-brain slice obtained using optical imaging at normal ambient pressure (upper panel) and at high pressure (lower panel). The amplitude of the signal indicates the degree of activity. Under high-pressure conditions, the assembly is much more extensive. A time course of the response from a small area of tissue is shown for each pressure on the right-hand side.[40] (To see this scan in colour, see Plate 3.)

been able to compare lighter levels of surgical anaesthesia with much deeper ones. It turns out that the rat brain generates an assembly significantly more extensive under lightly anaesthetized conditions compared to when the anaesthetic is deepened.[41] As we increase the anaesthesia level, so we reduce the size of the assembly in its cortex.

Moreover, the assembly generated under light anaesthetic is quickly curtailed by a rebound window of inhibited activity (Fig. 4), perhaps to ensure that the assembly is much more conspicuous as a clear signal against an unambiguous background. This effect is not seen once anaesthesia is deepened, most probably because it requires holistic brain interconnections to be operational.

Finally, although perhaps predictably, the speed it takes the assembly to be generated was faster under light anaesthetic conditions, once again there was an effect also on duration – just as we saw in the experiments on slices. With deeper anaesthesia, an assembly, evoked by whisker deflection, was now sustained for much longer: again, just as we saw in brain slices, but this time in the intact cortex, the anaesthesia seems to prolong an assembly, making it last significantly longer, which could in turn block any further assemblies forming.

So, we can conclude that neuronal assemblies are somehow linked to consciousness and its loss, for the following reasons: firstly, they are more affected by anaesthetics than by analgesics[42]; secondly, they are sensitive to the same conditions – high pressure – that reverse anaesthesia;[43] and finally, they reflect varying depths of anaesthesia.[44] Trying to piece together these three very different experimental findings to account for the action of anaesthetics would be premature just now: let's wait until we have collected more observations throughout the day before attempting any overarching theory. For the time being, however, we can at least see assemblies at work under various conditions. They are a true, naturally occurring phenomenon and, given their fast, transitory time frames, extensive spatial spread, continuous variability and sensitivity to consciousness-modifying treatments, they tick several of the boxes on our shopping list. Then again, it is of course no proof at all that this level of brain processing might indeed serve as a good index for measuring degrees of consciousness.

The dynamics of assemblies and effects of anaesthetics might correlate well, but that is a far cry from showing that they are linked

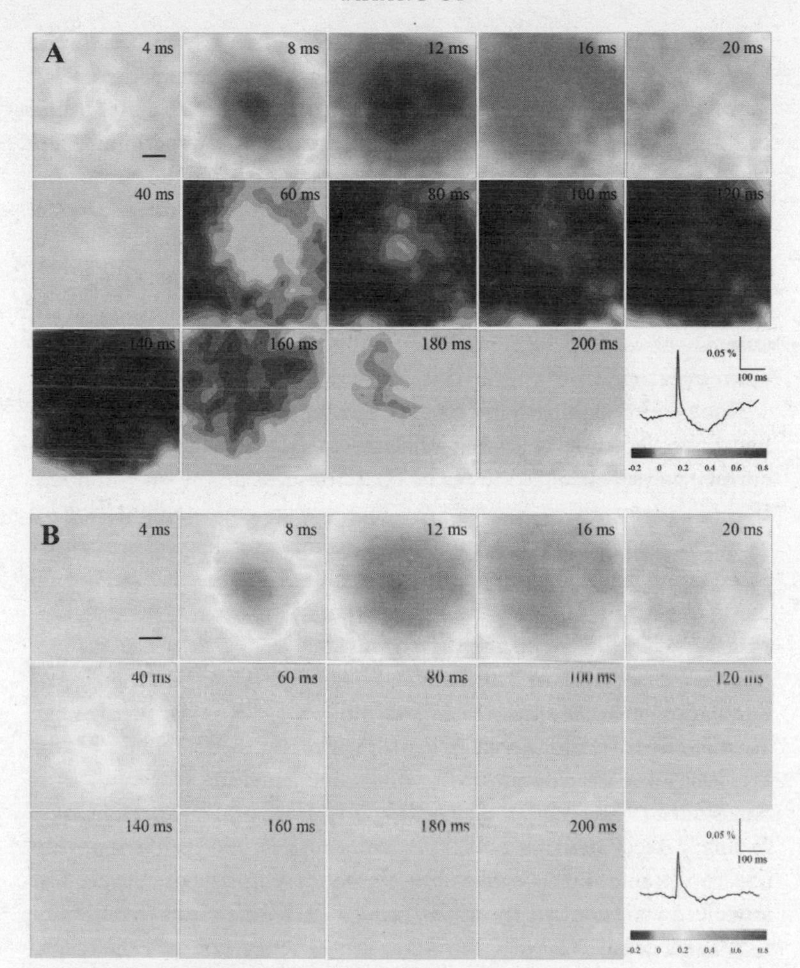

Fig. 4: Assemblies evoked by whisker stimulation in the rat during light (A) and deep (B) levels of anaesthesia.[45] Note that an inhibition occurs as a rebound after the intense activation during light anaesthesia but not after the reduced activation in deeper anaesthesia. Time after stimulus onset is displayed in the upper-right corners; note the change in time intervals in the second and third rows. Scale bar: 500um. (To see these scans in colour, see Plate 4.)

to determining whether or not we are conscious. Nonetheless, the anaesthetist Brian Pollard and his team at Manchester University[46] have independently developed a technique that further validates a link of some sort between assemblies and consciousness: Functional Electrical Impedance Tomography by Evoke Response (fEITER) electrically stimulates the brain, then monitors the resulting effect (electrical impedance) through the skull, with a time resolution twice as fast as VSDI, namely some 500 microseconds![47] While optical imaging with voltage-sensitive dyes will never be possible in conscious – or, indeed, even unconscious – humans, Brian Pollard's methodology opens up new opportunities for visualizing assemblies in humans.

Just to be crystal clear: what I'm suggesting is that transient configurations of large-scale neuronal assemblies throughout the brain correlate with varying degrees of consciousness at any one moment. If this theory proves correct, the monitoring and manipulation of assemblies would provide a valuable way of investigating previously intractable mental realities by eventually studying their neuronal correlates over space and time scales that for the first time are commensurate with those in the real brain.

Think of degrees of consciousness as being a bit like the ripples emanating from the throw of a stone into a puddle. Imagine throwing your stone, albeit into a puddle that's already made choppy by the prevailing breeze. Despite this, your stone generates clearly distinct patterns of ripples: the stone itself is permanent or quasi-permanent – in any case, a definite and fixed object. It is relatively small, but the ripples that it now evokes are disproportionately very large. The stone is quasi-permanent but the ripples are highly transient.

What is going to be very helpful to us is that the extent of the spread of ripples – that is, the extent of an assembly and thus the extent of consciousness at any one time – will be determined by a range of different, independent factors. For example, the ensuing assemblies (the ripples) will vary in size from one moment to the next according to the strength of the trigger – let's say, the alarm clock – and also according to the ease with which the neurons will be synchronized, which is in turn dependent on the availability of facilitating 'modulatory' chemicals that relate to arousal levels and sleep–wake cycles. But we don't go through life with our consciousness driven just by our raw senses. There will also be other factors that determine the eventual extent of an assembly at any one time.

Most importantly, there is the stone itself: as well as the external force with which it will be thrown – namely, the strength of psychophysical stimulation – there is the variability in size, which we could view as the extent of an internal, pre-existing, hard-wired network that is initially activated. We've seen that such a localized hub of functionally connected cells is standard fare in neuroscience: indeed, it was first mooted by Donald Hebb in the middle of the last century, and later validated by the phenomenon of 'long-term potentiation',[48] whereby cells that are active simultaneously form localized, enduring, hard-wired connections which we could now envisage as a kind of neuronal stone. Then again, on its own, such a hub of connected brain cells would be too local, too permanently wired up, and indeed too slow to form to accommodate our shopping list for consciousness.

Nonetheless, it is just such localized and enduring connectivity that would fit the bill as a 'stone' of varying size, just as its activation by incoming stimuli – say, the sound of an alarm clock – would fit the bill for activating it. What then would be the phenomenological equivalent of a large or small stone, an extensive or modestly hard-wired hub? As we track on through your day, we can start to find out. Outside, the sky is now bleached bright daytime white: it is time to get up.

3
Walking the Dog

As you open your eyes and stare around your bedroom, your immediate next thought is what you are supposed to be doing once you get out of bed. By now, at least, you are fully aware of who you are, where you are, what you did yesterday and what day it is today. And it's not just the forthcoming twelve hours or so, but the entire week, the unique fabric of your life, that is now stretching out reassuringly into a well-defined and well-arranged perspective. What could have happened in your brain to progress from the simple, passive consciousness of snuggling for a precious moment longer under the duvet to such inwardly directed focus?

The immediacy of the bedroom is giving way to a turbulent, hypothetical world of events and meetings ahead, and the part that you, the individual, will play in it . . . Your consciousness has undoubtedly 'deepened' or 'grown' in some way over the last few moments: some additional factor must therefore now be coming into play to contribute to this more meaningful and individual take on the world, something that has led you from the purely sensational indulgence of the here and now to this much more personalized 'cognitive' perspective.

You pull yourself up and out of bed, drag on some clothes and stumble down the stairs. Bobo, your border collie, is waiting impatiently and ludicrously ecstatic at this early hour, the lead in his mouth, tail flashing from side to side. There is no time for a coffee right now, let alone breakfast: you will have no peace until the two of you have embarked immediately on the routine morning walk. But perhaps the fresh air and the rhythmic pace of mechanically putting one foot in front of the other will give you time to reflect and think more deeply. As Friedrich Nietzsche once avowed, 'All truly great thoughts are conceived by walking.'

This suggestion might seem counterintuitive: surely the brain has limited resources and, when two tasks are carried out simultaneously – say, walking and thinking – then the performance of one or the other might be impaired: indeed, this very point has been proved most recently with high intensity of digital-media multitasking being linked to a reduced connectivity in certain brain regions.[1] Then again, an alternative view is that, unlike parallel-screen activities which are similar and compete for the same neuronal resources, there could be multiple pools of mental resources, each for very different tasks: when activities are completely distinct from each other there may even be a *synergistic* effect. It turns out that this could well be the case. One of the lesser-known benefits of walking in natural environments can be to improve attention: so cognitive performance is indeed improved when engaged in this other, completely different, behaviour of putting one foot in front of the other. Meanwhile, research shows that working memory is significantly better when participants walk at a speed of their own choosing.[2]

Moreover, immersion in natural rather than urban environments can make a big difference when it comes to how well you use your brain. Investigators based in Ann Arbor, Michigan, looked at two measures for evaluating cerebral power.[3] First, in order to assess attention, a task was chosen in which participants listened to a series of numbers three to nine digits in length, then repeated them back in reverse order: here, performance depends on directed-attention abilities because you have to move items in and out of your attentional focus. This is a major component of short-term memory. Second, a computer-based task measured three types of attention: alerting, orienting and executive, with executive attention requiring the greatest control.

Using these measures of mental ability, the idea was then to test the effects before and after a fifty-minute walk of just under three miles in the Ann Arbor arboretum, an area, as the name implies, that is dominated by trees and free from traffic. The effects of this environment were contrasted to the participants' experience of downtown

Ann Arbor, which, inevitably, involved exposure to a busy, noisy environment with heavy traffic. More fascinating still, the same tests were given to a separate group of participants before and after simply viewing either urban or natural photographs for ten minutes or so. The rationale for using photographs as part of the study was to add a control for the key factor in the apparently beneficial power of nature being simply due to the unusual quiet and peacefulness rather than the environment itself. Amazingly enough, the results showed that the experience of walking in natural environments, or even merely viewing photographs for a short period, had an impressive effect on brain function.[4] For those who took the arboretum walk or even just viewed photographs of natural settings, test scores significantly improved for executive attention: but the same could not be said for those undertaking the downtown walk or looking at photographs of urban environments. It seems that exposure even to two-dimensional rural environments specifically by virtue of the associations they trigger, will boost cognitive capacity: the total amount of information the brain is capable of retaining at any particular moment.

The researchers explain these findings by contrasting reactive with proactive attention: urban environments, and indeed most of our normal working lives, are full of dramatic stimuli that attract our attention, whether we like it or not. This type of environment thereby monopolizes still more reactive attentional processes so that we can avoid, say, being hit by a car. On the other hand, a natural environment spares us this relentless reactivity to the outside world and instead shapes our consciousness more subtly in a more voluntary, proactive manner. It is *you* who decides to examine a plant more closely or to focus on the far-flung horizon one moment, then perhaps to lean up against a tree the next: this internally driven sequence of events will then have the additional benefit of restoring a sense of control, of giving you a longer time frame in which to develop and deepen your thoughts.

A somewhat more extreme version of experimenting with the impact of nature involved a study of the effect of four-day hiking trips in Maine, Colorado or Alaska not just on cognition but on the still more rarefied phenomenon of creative ability.[5] All electronic devices were banned. Participants took a Remote Associates Test, a means of objectively evaluating creativity: the subject had to answer, typically, some thirty to forty questions, each of which consists of three everyday words that at first appear to be unrelated. The task is to think of a fourth word that is somehow linked to each of the first three. For example, the word in common with 'broken', 'clear' and 'eye' would be, for example, 'glass'. After a four-day hike, incredibly, the participants' scores of creativity and problem-solving were 50 per cent higher than for those taking the test before their hiking trip. However, due to the extreme, blanket nature of the intervention, it is difficult to pick out which of the many uncontrolled-for factors had the greatest impact – be it exercise, the all-pervasive natural environment, the absence of electronic devices or even just better sleep, fresh air and more social interaction.

In any case, this is the same kind of outdoor environment within which currently you are immersed as you stumble after Bobo as he strains on his lead. For now at least, you don't have to react urgently to incoming, fragmented bits of distraction ceaselessly alerting you to stimuli in the external world: instead, you have a longer time frame in which to digest the sensory inputs, piece them all together, even perhaps to join up the mental dots in a new, 'creative' way. In any case, you're starting to breathe deeply in the open air, to look not just outwards but inwards – to think.

When it comes down to it, a feeling is just that: a momentary experience in the here and now. It is not dependent on a start and a finish, on any previous background or on any future planning: laughing and screaming are spontaneous, atavistic responses. So the tail-wagging dog and the gurgling baby are respectively generating an evolutionary (phylogenetic) and a developmental (ontogenetic) expression of the most primitive type of consciousness, one dominated by raw sensation and reactivity. In contrast, now reflect on as diverse a range of different types of thoughts as you possibly can – a memory, a fantasy, a logical argument, a business plan: they all have an essential, basic feature in common which a raw feeling, an emotion, lacks. That crucial issue for a thought is surely the time frame: it has a beginning, a middle and an end at which, in all cases, you fetch up in a different place from where you started. And how did you get to this new place, this new conclusion, punch-line or solution? Via a chain of steps, in which one leads on to the next in a clear, linear path: we even talk of 'thinking straight'.

Tellingly, along the same 'lines' (the pun *is* intended), the distinguished neurologist Oleh Hornykiewicz, who developed the still-current treatment for the movement disorder Parkinson's disease, defined thinking as 'movement confined to the brain'. So if a thought, unlike an emotion, is a series of steps, then that would be a kind of movement: the longer the journey, the 'deeper' the thought. Moreover, the actual physical act of walking could amplify and thereby perhaps enhance this inner process: by reflecting in external movement what is happening in the brain, by having a clear causal link between one step and the next, with the mental being enforced by the physical, the repetitive contraction of muscles could help insure against the mind 'wandering', going, literally, off track.

And walking with a dog might enhance the process still further . . . As the political journalist Ed Stourton recorded in his delightful *Diary of a Dog Walker*, 'Dog-walking becomes like reading a novel, or watching a play: disbelief is suspended and, for an hour or so, we are given licence to escape ordinary life. Fantasy flourishes and really quite trivial moments in a dog's life become a source of wonder to be laughed about and even worried over.'[6]

Any dog walker would probably insist that their dog was conscious – why, otherwise, lavish so much time and money on your pet? As Ed Stourton puts it, 'While we may know that a dog is just a dog, we continue to allow ourselves to speak and think of our dogs as friends, individuals with a full claim on our affections.' But how exactly *is* a dog conscious, and how might such a subjective canine experience differ from that of our own species?

NON-HUMAN CONSCIOUSNESS

This is really the same basic question as the one that cropped up in our discussion of foetal consciousness: when do all the lights go on inside the newly minted human head? Perhaps these twin riddles, posed by the foetus and animals alike, give us a clue. So let's think a bit more about non-human versus human consciousness in the same way as we've already considered foetal consciousness in the previous chapter: just as consciousness might grow as an individual grows (ontogenetically), mightn't the very same process be enacted on an evolutionary (phylogenetic) stage? In other words, could consciousness develop in parallel with evolution?

The human brain is different from that of the rat, the squirrel, the rabbit, the camel, the cat, the monkey, and so on, just as each respective species is different from another. In the animal kingdom, each species has a characteristic brain structure with its own variations, in which some anatomical regions might prosper at the expense of others, some may be exaggerated in shape and the overall size of the brain will vary enormously. But the basic structural blueprint is the same: there is no clear anatomical line, no Rubicon you can cross where you could be confident enough to say, 'This species is conscious. That species is just an insensate machine.'

Just consider for a moment the following example flagged by Toby Collins, a marine biologist who became fascinated by neuroscience and ended up working in my lab. Because of the unusual perspective he had from his earlier professional incarnation, Toby raised some interesting questions about the octopus. In many countries around the world, government legislation is in force to regulate experiments on living animals – most typically, vertebrates – to ensure that the procedures are as humane and pain-free as possible. Interestingly enough, in the UK at least, the list of species protected in this way also includes an invertebrate, the *octopus vulgaris*: you cannot perform invasive procedures such as surgery without anaesthetic, and so the assumption must be that it feels pain. However, the *eledone cirrhosa*, a very closely related octopus species, benefits from no such legislation, so would merit no comparable consideration. Yet where does the distinction lie? There is no real, significant anatomical differential between these two types of octopus and yet, paradoxically, one is protected legally, while the other is not. There is obviously no clear line where we could say, with unequivocal confidence, that one animal is capable of an inner subjective state yet the other is merely some kind of eight-tentacled machine.

Remember the World Cup of 2010, when the celebrity octopus, Paul, was apparently so 'conscious' he could predict the scores of upcoming matches? Nonetheless, cephalopods such as Paul have never really been evaluated formally, regarding their 'intelligence', compared with other closely related species: however, they have long been used in memory experiments, thanks to the focus placed on them by a talented young marine biologist working in Naples in the 1920s, J. Z. Young, who was fascinated by their anatomy. Young was the first to show that the octopus is indeed capable of significant cognitive processes: for example, they could discriminate between different objects based on size, shape and colour.[7] Moreover, the manner in which octopuses classify different-shaped objects turns out to be the same as that employed by vertebrates such as goldfish and rats.[8]

More recently, octopuses have demonstrated their memory skills and problem-solving capabilities by learning the correct path through a Plexiglass maze to a reward, or by retrieving an object from a clear bottle sealed with a plug.[9] More widely, as a group, cephalopods appear to be capable of an impressive behavioural repertoire of adaptation, while anecdotal reports of apparent learning indicate highly developed attention and memory capacities. And in experiments where an octopus was challenged with a maze containing obstacles, the scientists went so far as to suggest that the octopus appeared to 'consider' the layout of the maze before proceeding.[10] This clearly intellectual cephalopod can even solve problems through observational learning, so not just through mimicry but instead in a way that implies real memory skills.[11]

The surprising mental prowess of these invertebrates suggests that the octopus in particular, and cephalopods in general, should have as good a claim to consciousness as any other species in the animal kingdom. If they do, however, then we're faced with a truly interesting riddle. The brain of the octopus bears no similarity to that of mammals, and the anatomical areas such as the cortex and the thalamocortical loop, which have been considered so vital for consciousness, simply do not exist in any comparable form.

The brain of an adult octopus is made up of 170 million cells, most of which are neurons.[12] While this may sound impressive, the human brain boasts some 86 billion at least.[13] Nonetheless, cephalopods such as the octopus have sophisticated sense receptors and nervous systems that are comparable in complexity to those of some vertebrates, such as birds. Yet, while the cephalopod brain may not have precise and elaborate thalamocortical loops, it *does* have the more basic machinery of neurons and synapses, along with transmitters such as dopamine, noradrenaline and serotonin.[14] So, even though it has not yet been directly demonstrated, the arguably conscious octopus would still have all the wherewithal for generating assemblies. We've already seen that we should free up our thinking, away from the rigid framework of macro-anatomical regions, and that, in any case, 'key' brain areas can never provide any real explanation *per se* as to how consciousness is generated. Instead, the differences in non-mammalian brains prove yet again that assemblies will provide a more helpful starting point in identifying neural correlates of consciousness: after all, these meso-scale phenomena are much less anatomically and physiologically prescriptive for only certain types of animal brains. In addition, assemblies are analogue, and hence flexible, continuously expanding and contracting in size (and hence with the potential for generating variable consciousness) from one moment to next, from one stage of development to the next – and indeed from one species to the next.

It could well be that 'simple' animals such as the octopus are indeed conscious, but not as conscious as, for example, the rat. And the rat is conscious, but not as conscious as a dog or a cat. And dogs and cats are conscious, but not as conscious as primates – just as we saw that a foetus could be conscious, but not as conscious as a child; and a child could be conscious, but not as conscious as an adult. But then, where would the all-important *quantitative* evolutionary differential lie, in terms of degree of consciousness? If it is indeed, as for the octopus, the basic potential to form transient, large-scale neuronal coalitions, then perhaps, as with the riddle of the foetus in the previous chapter, the crucial issue is net assembly size. Let's now revisit the various parameters that will determine this final estimate for depth of consciousness – the extent of the ripples when a stone is thrown. This metaphor will be useful as we go along, as we can use it to unpack in turn the different factors contributing to the final outcome. The excursion of the ensuing ripples will vary enormously because stones can vary in size, as can the force with which they are thrown – and this analogy holds well when it comes to exploring brain processes. Now that we have a good shopping list, we can at last visit the brain and investigate what the neuronal 'stone' might be and what brain process is the equivalent of 'throwing' it.

MAKING RIPPLES: SIZE OF STONE AND FORCE OF THROW

Whoever you are, however young or old you may be, whatever kind of brain you have, whether you are a human or another species, unless you are stone deaf, an alarm clock going off will always wake you up. Why is that? Why should an unusually loud noise immediately push you into consciousness?

You can think of the alarm clock as the neuroscience equivalent of throwing a stone. Even a small stone, if thrown forcefully enough, can generate the all-important ripples: the noise (the throw) would be loud enough to activate through the normal sensory pathways a hub of hard-wired, auditory-related brain cells (the stone) that is so forceful that a further, far more extensive coalition of cells will be recruited, i. e. an assembly (the ripples) will be temporarily generated. Any animal with a functional auditory system, of any age or culture, will be yanked out of sleep and into some kind of consciousness by the loud noise. The type of consciousness that then results is another issue altogether and needn't worry us for now: the crucial factor here is simply the trigger – the size and force of throw of the stone as determinants of the final extent of the ripples.

While the force of the stone throw – the loudness of the alarm clock, say – is determined by external factors, the actual size of the stone will be dependent on the internal set-up of the individual brain in question. Within any species (but particularly marked in more sophisticated animals such as ourselves), the configuration and connections of individual hard-wired hubs of brain cells could be critical in determining the next important variable factor: the neural equivalent of the size of the stone. As our brains develop, so the connections will be shaped by our experiences of the outside world: this phenomenon, whereby experience almost literally leaves its mark on your brain, is, as we learned earlier, known as plasticity.

In the brains of all species, there will be neurons that are activated to varying strengths by the incoming senses (the variable force of the throw of the stone): but the number of neurons in a hard-wired hub (the size of the stone) will vary independently, according to the specific configuration and connectivity of neurons within an animal's brain. This means that the *same* sound or sight at the *same* intensity will have *different* effects in different brains as the respective size of the stone varies – that is, the number of hard-wired cells that can be activated. In turn, the extent of this hard-wired hub will depend on an additional, crucial factor: earlier personal interaction with the individual environment. The more sophisticated the species in question, the more its capacity for individual experience to have left its mark on the brain and the greater the variation in size of the stones. So, let's think again about the stone being thrown into a puddle: it doesn't have to be forcefully thrown – as in the brutal case of the alarm clock. Instead, extensive ripples could be generated simply because the stone is large; in neuro-speak, a hub of brain cells that has gradually expanded through persistent, hard-wired connections, driven in turn by individual experience exploiting brain plasticity.

This experience-driven growth in connectivity, and hence in the larger size of the stone, can also occur, albeit to a lesser extent, in non-humans. Take adult rats, for example. A rat may attach no enduring personal significance to its own mother, but the particular environmental experiences of its surroundings more generally will nonetheless leave their mark on its brain. A popular way to examine this effect in the lab is by creating a so-called 'enriched environment'. 'Enrichment' for a rat doesn't mean they need to eat exotic food from bejewelled hoppers or live in golden cages. Instead, 'enrichment' refers to an environment that stimulates the brain as much as possible: so, if you want to give a rat maximal stimulation, what you do is ensure that it has opportunities for highly exploratory experiences with diverse novel objects (see Figure 5).

Despite their widespread demonization in both fact and fiction, rats are in fact very curious and intelligent creatures and will, if given half the chance, embark on a full investigation of wherever they are. Correspondingly, their brains will reflect this lifestyle. The first empirical demonstration of such *experience-dependent* plasticity in an enriched environment was in the 1940s: Donald Hebb, whom we met earlier, took lab rats home and let them thereby experience a novel, interactive environment very different from their usual cages at the lab. After some weeks in Hebb's house, these 'free-range' rats exhibited superior problem-solving ability, compared to their less lucky lab counterparts who had remained in their usual cages.

However, the palpable fact that there are actual physical changes in neuronal circuits as a result of stimulation was demonstrated directly only several decades later.[15] The scientists had actually set out to identify the neural mechanisms underlying individual differences in behaviour and problem-solving in various strains of rats but quickly realized the enormous influence of heightened experience.

As you can imagine, the impact of environmental enrichment has been the focus of much attention from neuroscientists and psychologists: we now know that it has an impact in a wide range of species, and at all ages. Studies over the last few decades on 'enriched' animals have shown clear anatomical changes, all of them positive:[16] both young and aged subjects are superior in tests of spatial memory, while even a brief period of environmental enrichment rescues adult memory deficits in a certain strain of genetically modified mice, used as an animal model of Alzheimer's disease:[17] moreover, this kind of

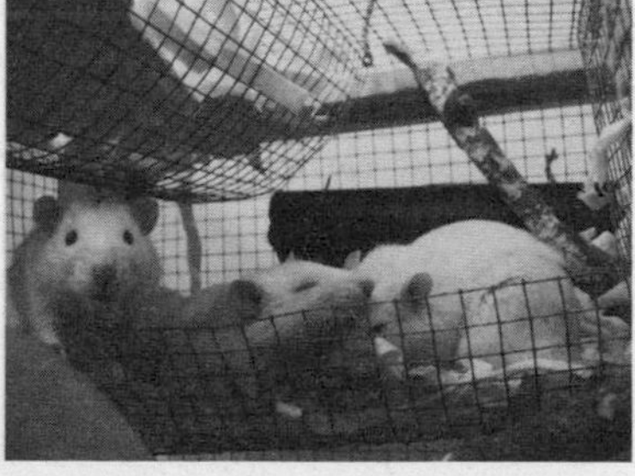

Fig. 5: A typical 'enriched' environment for rats (from Devonshire, Dommett & Greenfield, unpublished).

stimulation induces neurogenesis (the production of new brain cells) and improvements in memory.[18]

Environmental enrichment, even in rats, can also help delay or mitigate brain damage more generally. For example, Huntington's disease is an inherited, single-gene disorder in which there is progressive neurodegeneration and for which there is no known cure. 'Transgenic' (genetically modified) mice have been investigated which develop a neurodegenerative syndrome closely reflecting the escalating movement difficulties associated with the human disease over time: however, exposure of these mice to a stimulating, enriched environment from an early age helps to prevent the loss of brain tissue and delays the onset of motor disorders[19] as well as offsetting brain damage.[20]

A crucial factor here is the duration of the enrichment experience. For example, one study investigated the way in which different periods of enrichment experience affected mouse behaviour – specifically, their locomotion – over periods of exposure of one, four and eight weeks. One week of environmental enrichment did nothing, but four weeks' worth had behavioural effects that lasted two months, and eight weeks set in train effects lasting six months. Clearly, a minimum duration of enrichment is required in order to see structural and functional effects, and these consequences can extend beyond the actual period of the experience, with the persistence of effects related directly to the original duration of enrichment.[21]

All of these studies illustrate the importance of a basic key factor: the interaction of an animal, irrespective of species and its respective cognitive prowess, within a stimulating environmental scenario. Similar environment-induced changes have now been found in mice, gerbils, squirrels, cats, monkeys and even in birds, fish, fruit-flies and spiders – in short, in every animal 'from flies to philosophers'.[22] It is not just passive exposure to a novel scene that is important but how the subject in question – more specifically, the particular brain – responds to it.

The all-important interactive environment makes brain cells work harder so that, as with the exercising of muscles, they grow. However, if you look at a single brain cell in animals exposed to enrichment, compared to those kept in ordinary old lab cages, then you can see that brain cells are not quite the same as muscles in the way they respond. Our brain cells don't simply get bigger in the way muscles do; instead, they grow more branches, known as dendrites.[23] While rodents have been the most commonly tested species, in primates, too, the environment causes enhanced structural and chemical changes including greater branching of neurons.[24] So why is this particularly interesting or important? Because, by increasing its branches, a brain cell is increasing its surface area, which means that it can make more connections.

Inevitably, these types of highly controlled environmental-enrichment paradigms can be systematically investigated only in non-humans; it is hard to imagine a similarly controlled environment for people. Still, this experience-dependent increase in brain-cell connectivity could have important implications for our own species. As the brain becomes more sophisticated across the animal kingdom, the significance of personalized connections will be important for a unique shaping of neuronal configurations within the brain, bestowing on an individual their unique take on the world.[25] We know that, when the human brain grows in the early years of life, this increase is not due to the proliferation in number of actual brain cells but more to the growth of the connections between them.

In the human cortex, the number of synapses (the connections between the neurons) increases rapidly in prenatal life and continues to grow for a brief period post-natally. Thereafter, the number of synapses slowly falls off over a long period of time, from the age of approximately six months through to adolescence, to reach a stable level in adulthood.[26] In an imaging study on human subjects between the ages of four and twenty years old,[27] it was seen that the cortex increased in volume during brain development up until the age of around ten or eleven in boys, and eight or nine in girls, before slowly decreasing as a result of synaptic 'pruning': accordingly, the brain then becomes less unconditionally open to any possibility and more tailored to meet an individual's needs.[28]

In adults, there are many examples of human brain plasticity, whereby neuronal connectivity is still incessantly being shaped by the environment, illustrating over and over, in many varied situations, how structural changes in the brain clearly do not just accompany the learning of new skills but reflect daily life itself. There are two broad types of study using human subjects that reveal this astonishing talent of our brains.

The first type are 'snapshot studies', like the one we saw previously of the London taxi drivers,[29] where, in a single one-off observation, professionals or experts engaged for unusually intensive and long periods in a particular type of behaviour show marked differences when compared with people not so expert or focused. Take mathematicians, for example: the prolonged time they spend engaged in maths induces increased cell density in an area of cortex (the parietal lobe), that is involved in either arithmetic or visual processing. Then there are musicians, in whom brain structures can differ conspicuously in comparison to non-musicians. Brain scans of professional keyboard players, amateurs and non-musicians have revealed grey-matter increases in a range of motor, auditory, and visual-spatial brain regions. Moreover, there are strong links between a musician's ability and the intensity of their practice, suggesting that these anatomical differences are related to the process of learning itself and not to some innate predisposition to music.[30]

Extensive piano practice, in particular, has specific effects in certain regions of the brain on development of neuronal connections (white matter) in subjects of all ages – children, adolescents and adults. As you might expect, the developing brains of children show greater levels of plasticity. But other age groups continue to show effects, though this time only in the cortex: clearly, specific time windows ('critical periods'), play a role in regional plasticity of certain areas of the central nervous system.[31]

Given the wide-ranging diversity of skills such as taxi driving, playing a musical instrument and spending time on maths problems, it might by now be unsurprising that other wildly diverse activities can also leave their mark on the brain. Take golf: in one survey of brain scans of skilled and less-skilled golfers, and non-golfers, larger grey-matter volumes in particular brain structures were revealed only in the skilled players.[32] And there's nothing unique about golf as a sport in terms of brain adaptability. Experience-dependent plasticity is also detectable in the brains of basketball players, this time as an enlargement in an area known as the 'autopilot' of the brain (cerebellum), which is related to tightly coordinated sensory-motor interplay and is just the type of complex motor performance seen in professional players.[33]

Then there is another type of experiment altogether. Instead of just taking a one-off snapshot of the brains of dedicated experts, compared with ordinary mortals, the second, alternative way of investigating human brain plasticity is with longitudinal studies. This time, the experiment takes place with multiple testing over a defined period of time so that you can see before and after effects of a particular experience within the same 'average' subject: the participants would be ordinary people and start off initially with no special skills, but during the experiment they would learn a specific activity or expertise.

For instance, it turns out that learning a language drives plasticity in the brain, increasing the density of grey matter; the changes observed correspond to the level of skill in the language. In another study, brain plasticity in the language system was linked to an increased second-language proficiency: five months of second-language learning, by native English-speaking exchange students learning German in Switzerland, results in structural changes (in the left inferior frontal gyrus) that parallel an increase in proficiency. Once again, it is the process of learning that counts, and which is reflected in changes in the structure of the brain.[34]

In a very different project, subjects were asked to learn a three-ball juggling task, in which perception and anticipation are key to determining accurately upcoming movements. The volunteers underwent daily training for a three-month period: scans were performed before any training, after three months, and once again after six months, there having been no juggling between three and six months. After three months of juggling, there was an increase in the size of a particular part of the cortex. Yet this change was short-lived. Performance was much worse at the six-month scan, and structural changes had almost returned to baseline. A follow-up study demonstrated that the brain adaptations had already occurred within seven days of the start of training. Because such transformations are most rapid during the early stages of the learning process, when performance level is low, it seems that it is the actual act of acquiring a new task that is pivotal in changing the structure of the brain, rather than the ongoing practice of something already learned.[35]

But it's not just formal learning and physical activity that can have consequences for neuronal connections. One of the most intriguing studies of plasticity ever was based on the effects of piano playing. Over the course of only five days, as subjects learned a one-handed, five-finger exercise through daily two-hour sessions, imaging studies showed that the areas of cortex targeting the appropriate hand muscles had expanded and their activation threshold was lower: no changes occurred in control subjects who underwent daily mapping but did not practise on the piano at all.[36] These findings are similar to those of the piano study mentioned previously, but this particular investigation went one step further. Most astonishing of all, such changes in the brain were also observed with just 'mental' practice, where a different group of subjects were asked to imagine they were playing the piano without actually doing so. Surely then, this result shows that you cannot talk about 'mental' as the antithesis of 'physical': that old dichotomy just doesn't make sense any more and it isn't helpful in asking and answering the big questions relating to the mind and consciousness.

The other exciting implication of the finding in this study is that, in terms of brain function, what is important for plasticity is not the actual contraction of the physical muscle: instead, it is the thought that has preceded it. Oleh Hornykiewitz's observation that 'Thinking is movement confined to the brain,' is exactly what these brain scans are surely showing.

There are many conclusions we can draw from all these explorations of plasticity, but perhaps the two most important are that mental and physical activity alike leaves its mark on the brain, and that this plasticity is not exclusive to just one special set of neurons but seems to be a generic phenomenon occurring all over the brain. To greater or lesser extents, all brains, even that of a sea slug,[37] adapt to experience, but our species outperforms any other. The close interdependence of Nature and Nurture that we now know characterizes much of biology[38] tilts strongly towards the environment when it comes to developing human mental functions and an individual mind.

This is a remarkable notion: no one in the past, for the hundred thousand years our species has been around, has had a brain or, more accurately, a mind, like yours – nor could they ever again. Human beings, as they develop, live their lives less and less in a state in which every conscious moment is triggered by purely sensory stimulations needing an immediate, often far more stereotyped reaction. As each of us develops, there is a growing tendency for a cognitive experience – one that goes beyond the face value of encounters with the various objects, people and events of daily life. As a result, we are freed up from purely raw sensory perceptions in order to have a more meaningful and personalized take on the world. How might this be achieved?

DEVELOPING A MIND

In the words of William James, considered by many to be the father of American psychology, you are born as a baby into a 'booming, buzzing confusion'. But gradually, as the weeks turn to months and these convert into years, erstwhile abstract colours, shapes, textures and smells coalesce into something that is, say, your mother's face. As Mum features again and again in your life, she becomes 'significant' to you, and the basis of this significance is that more and more associations will be formed. As with the examples of plasticity, your brain will adapt with a unique configuration of brain-cell connections: your mother will come to 'mean' something to you in a way that is completely unique, even among your siblings. It's through these highly personalized connections that you have a way of escaping from the press of your raw senses into a more cognitive state of seeing, literally, beyond face value.

Slowly, you progress from there being a one-way street of sensory bombardment to a two-way dialogue between your brain and the outside world. Now, instead of being constantly in a confusion of abstract colours and sounds, incoming stimuli – a person, an object, an event – will 'mean' something. You will evaluate them in terms of the existing neuronal connectivity while, at the same time, the very experience of so doing will update further the status of those neuronal connections. Now you can recognize your mum by virtue of the associations, the connections, between your particular neurons, which are unique to you and your brain.

A wedding ring, to take another example, might perhaps first be of interest to a small baby simply because of its sensory properties – the gold gleam, the central hole, the smooth, round surface that rolls along a surface. But as connections associated with the ring gradually become established, so the object will slowly gain meaning as a particular type of object, further defined eventually as something you put on one particular finger and only under particular circumstances. Then, eventually, if you acquire a wedding ring of your own, that specific object will have a specific meaning beyond its financial value, a relevance that any other very similar-looking ring does not possess, thanks to the personalized experiences and hence unique neuronal connections now operating in your brain alone. It is these connections between your neurons, and nothing to do with the intrinsic properties of the physical object itself, that give this specific ring a deep, 'special' significance, even though, in purely sensory terms, it would be unexceptional.

The difference between any old wedding ring and the particular one that might be, or once was, the most important object in your life, is one played out entirely in your head. In this way, the formerly one-way street now has traffic going in both directions. Everything you are currently experiencing from one moment to the next is read off against pre-existing associations: but, at the same time, that current, ongoing experience will be updating the connectivity, to change it for ever. As you grow, the development of your 'mind' is characterized by this increasingly vigorous and unique dialogue between your brain and the outside world.

The human 'mind' then, far from being some ethereal and exotic alternative to the everyday squalor of the physical brain, could instead be viewed as the personalization of the brain through the unique configuration of neuronal connections, driven in turn by an individual's unique experience.[39] The Halle Berry neuron we came across earlier could be activated by non-visual, purely cognitive associations with the actress, which would be part of just such a hard-wired hub. Your brain is unique because you have unique neuronal connections, and these networks reflect your individual experience. So what you have experienced will determine how you see the world: there is a constant interaction between how you evaluate everything around you in terms of pre-existing networks while, simultaneously, the ongoing current experience will be constantly upgrading that very connectivity.

The actual neuronal mechanism underlying this hard-wired connectivity – a process known as 'long-term potentiation' (LTP) or 'long-term depression' (LTD) – has been worked out in step-by-step detail by neuroscientists over several decades:[40] however, the careful cascade of this now well-established biochemical event is too slow to form, too long-lasting and too local, to accommodate a moment of consciousness – or indeed a neuronal assembly.

However, the 'mind' is not synonymous with consciousness, as we deduced in Chapter 1 – the two can be dissociated, as either can be operational without the other. For example, during sleep, no one would claim you had lost your mind and, indeed, the personalized connectivity endures, while on occasions where you are 'out of your mind', you are still conscious, but the processing of the immediate environment and/or ongoing behaviour is not dependent on the full extent of connections. Yet, most typically, for the everyday subjective experience of the adult human, this mind, though distinguishable from consciousness, will be playing a part in it: the particular size of the stone will be an important factor in determining the eventual one-off excursion of the ripples – the momentary degree of consciousness.

To summarize: a combination of the *force* of the throw, namely, the degree of activation (in phenomenological terms, raw sensory stimulation) and its *size*, the hard-wired local neuronal connectivity (in phenomenological terms, personalized 'meaning') will determine the overall effectiveness of the stone throw, that is, the extent of the ripples, or the size of the assembly (in phenomenological terms, the degree of consciousness at any one moment). This scheme is shown in

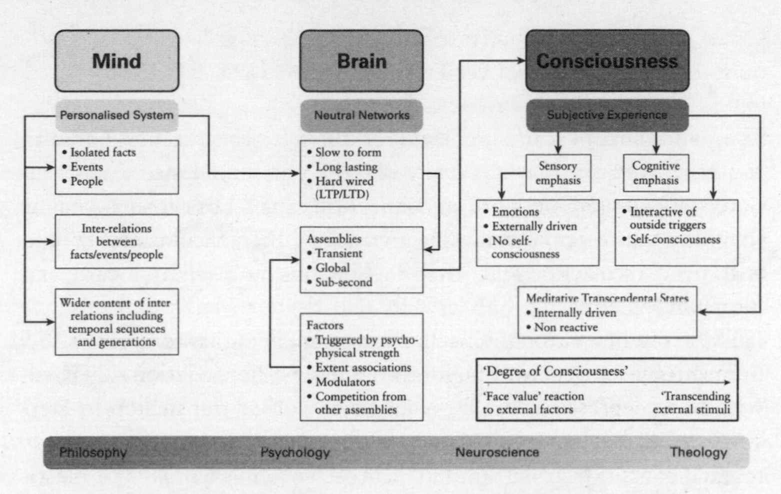

Fig. 6: 'Mind', 'Brain' and 'Consciousness': 'Mind' is the personalization of the brain through the adaptive plasticity of local, long-lasting connections between brain cells (see 'Brain') which can, nonetheless, be expressed in phenomenological terminology relating to facts, events and people. Within the brain, however, the corresponding 'hard-wired' hubs can, if activated, and in the presence of potent modulators, then trigger a much more global and transient assembly, the extent of which will correlate with consciousness of varying degrees (see 'Consciousness'). In turn, degrees of consciousness will be related to the relevant contributions of external sensory inputs and of internalized cognitive ones. The most basic consciousness, shared by infants and non-humans, will be sensory driven, while the most sophisticated – the exclusively internally driven cognitive states – would be exemplified in rarefied human behaviour such as meditation.

Figure 6, where the distinct entities of brain, mind and consciousness are nonetheless interconnected with neuronal assemblies as a pivotal process.

FROM MIND TO CONSCIOUSNESS

Imagine someone catching sight of their mother, a familiar and meaningful figure whose image will trigger an abundance of associations (an extensive network of connections), as a large stone. And, as they do so (the act of throwing the stone), then their visual system will spring into action and, after a while, this hard-wired, locally and persistently connected hub of cells will begin to buzz. But now we face the big question: how is this hub – this neural stone that's just been thrown – going to recruit other remote cells, not normally hard-wired, to generate a much more extensive, albeit transient assembly? The reason this question is so daunting is because we now need to make a transition from familiar neuroscience mechanisms of plasticity (the size of stone into the puddle) to seeing how the metaphorical ripples might be generated to correlate with varying degrees of consciousness itself.

The answer lies in the status at that particular moment of the powerful modulating chemicals: the family of molecules spraying over vast banks of brain in accordance with varying levels of arousal. Whichever chemicals happen to be available at the time will then sensitize, or modulate, the response of the surrounding cells to the activated hard-wired hub (the stone) into participating in an assembly (creating ripples). In turn, the ease with which all these many more cells are transiently recruited to work together will depend on this family of fountaining chemicals (in phenomenological terms, the prevailing arousal levels and mood at any one time): this, then, is yet another factor in generating assemblies. The variable availability of these modulators will be analogous to varying the viscosity (that is, the thickness) of the puddle water, which, if dense with algae, will be a less effective medium for the ripples to spread out than if it were fresh rainwater.

The degree of consciousness at any one moment, therefore the extent of the ripples, will be a product of 1) externally driven raw sensory activation (the force of the stone throw), 2) internally generated personalized significance (the size of stone), in turn reflected in the degree of long-term hard wired neuronal connectivity, as well as

3) general arousal levels, namely, the availability of certain fountaining modulators (the viscosity of the puddle water). If this is actually the case, then it will be possible to manipulate the relevant domination of any of these three factors differentially.

Take a scenario where the stone is relatively small in size. As with the alarm clock, the raw sensory stimulation would have to be sufficient to recruit an assembly – albeit a relatively small one. Imagine, for example, you are dancing in a club in Ibiza to the insistent beat of music and phantasmagorical light displays. Your carefully constructed individual connectivity will be neither relevant, nor needed, during such a 'sensational' experience. A second way to disable your personalized neuronal connections would be to take psychoactive substances, which could in any case be part of the rave experience: drugs impair the synaptic efficiency of the hard-wired hub so that, functionally, it will be less effective.

For many, a fast route to pleasure is alcohol, which should also therefore reduce assembly size. Because it is highly fat soluble, meaning that it can easily penetrate the tight junctions of cells (the blood–brain barrier) separating the brain from the circulation system, alcohol enters quickly into the brain, where it acts on a particular transmitter we have not yet encountered: gaba-aminobutyric acid (GABA).[41] GABA will trigger the influx into your neurons of negatively charged chloride ions, making the inside of the cells correspondingly more negative (hyperpolarized): this means that the cells will be less able to generate action potentials – they will be inhibited. Perhaps it's not surprising, therefore. that GABA can indeed substantially reduce assembly size.[42]

A third option would be a comparable 'sensational' experience created if the stimuli flooding into your brain were extremely fast-paced. Fast-paced activities such as white-water rafting, skydiving and skiing can have a similar effect, but with a difference: although the stimulation may not be super-bright or unnaturally noisy, the fast pace of one event following another in rapid succession will mean that a fast volley of competing assemblies will prevent any input being sustained long enough to be processed and realized to its full extent.

In all three cases, the emphasis is tilted in favour of a small stone but one that is thrown forcefully. What is particularly interesting is that, although the activities may be very different, they have a crucial common factor: the feeling of being 'out of' your mind. You have now returned to the experience of being a small child, living for the moment as the passive, unconditional and non-self-conscious recipient of your senses. If this is the same type of here-and-now consciousness that infants and non-humans have, then we can correlate such rapidly forming and small assemblies with a subjective experience characterized by the type of unadulterated 'pure' emotions which infants and non-humans readily display. Most of the experiences generated by fast-paced sport, music, and so on, are recreational activities of choice, so they are, presumably, pleasurable. Perhaps we could extend this thought to link pleasure with unusually small assemblies in a more precise way, by looking at the chemical landscape of the brain and starting with a well-known transmitter: dopamine. So far, dopamine has featured only as part of a collective, a 'fountain' in the brain, where it is a close relation to other transmitters (noradrenaline and serotonin): but now it's worth a closer look in its own right.

The release of dopamine, in particular, has long had a close connection with pleasure: ever since pioneering experiments back in the 1950s showed that, instead of eating, rats would prefer to press a bar to stimulate electrically their own brain[43] – so long as the electrodes were implanted specifically in areas of the brain that released dopamine. Moreover, all recreational drugs have a final, common action within the brain: to release dopamine, which acts like a fountain in the brain, to access wide ranges of 'higher' centres. Raised levels of dopamine are linked to levels of heightened arousal and excitement: just look at the stimulant effects of the drug amphetamine, which triggers surges in this powerful molecule.

So, if dopamine is related to feelings of pleasure, and if small assemblies are also a common feature of pleasurable experiences, then one would think dopamine would reduce the size of assemblies. In some pilot experiments in the lab, this is just what we have found. Apomorphine, a drug which works as an impostor for dopamine in the brain, has a clear effect (see Figure 7) of reducing both the size, and also the duration, of the assembly in the cortex of a rat-brain slice.[44]

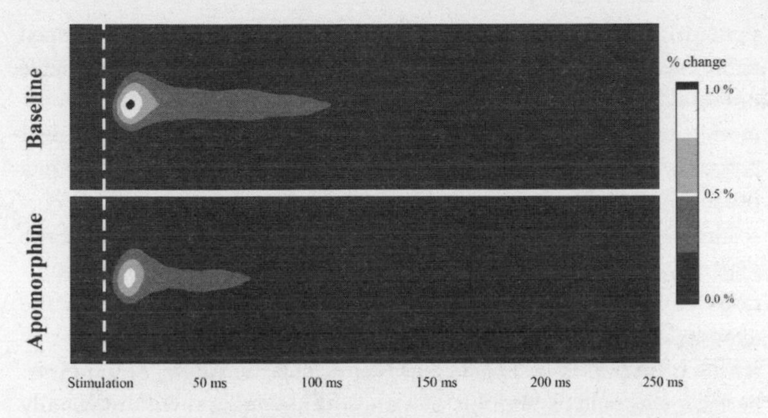

Fig. 7: An assembly generated in a rat-brain slice (cortex) over time. Note the reduction in intensity, as well as the strikingly shorter duration, when the dopamine impostor apomorphine (lower panel) is applied (Badin & Greenfield, unpublished).[45] (To see this scan in colour, see Plate 5.)

But a high level of dopamine need not always be the harbinger of pleasure: it can also play a part in the experience of fear. Take schizophrenia, a complex mental condition but one in which, among many other things, sufferers have an excessive level of functional dopamine. No one would claim that schizophrenia is fun. But, if you think about it, more generally wild and abandoned joy can frequently shade into fear: the game of peek-a-boo might scare a toddler, while on a roller coaster, screams are often very real. How might such polar opposite experiences, one positive and one so negative, shade into each other so readily?

In the case of pleasure, we've just seen that many recreational activities are 'sensational'. Yet, usually, the stand-alone, momentary hit of one moment will follow in the same genre as that of the previous one: the beat and cadences of the music are the same *type* of sensational experience, as are the successive and iterative touches of a lover, or the repetition of eating one savoury morsel after another. By complete contrast, when it comes to fear, psychologists have for a long time acknowledged that it can be related to anything highly novel, when, by definition, the completely unpredictable occurs.[46]

Perhaps pleasure turns to fear when we don't know what will happen next: each moment becomes a fragmented and utterly unpredictable one-off. This transition from pleasure to fear could actually be down to varying levels of dopamine and the opposite effects these different amounts can have on how the neuron will or will not participate in a potential assembly.[47]

However, pleasure, and its correlate of a small assembly, need not depend exclusively on moderate levels of this hard-working neurochemical. We saw earlier that consciousness-depriving anaesthetics also reduce assembly size, and they do so through stages, one of which is traditionally classified as mania: this 'delirium' would occur as the anaesthetic is taking effect and the assembly therefore already diminished. Tellingly, anaesthetics in doses too low to attain oblivion have traditionally given rise to pleasure. In the past, the familiar anaesthetic, ether, would have a big impact on those who were aiming not in any way to abolish their consciousness but instead to transform it: public gatherings, descriptively labelled 'ether frolics', would involve members of the audience inhaling diethyl ether to demonstrate the mind-altering properties of such agents. Similarly, the anaesthetic nitrous oxide was once used as a recreational drug at 'laughing gas' parties in the nineteenth century, at which some participants could become dreamy and sedated while others might enter into fits of giggles and experience a state of euphoria. As it happens, nitrous oxide has now made a comeback in the UK as a currently legal recreational drug, where it is inhaled, typically, from black balloons, understandably to the concern of the authorities.[48] The literally sensational experience of pure pleasure, whatever chemicals might be first driving it, seems consequently to correspond consistently to the high-emotion and low-cognition profile of a small assembly.

If the determining element in a sensational experience can be the next sip of wine, the next bar of music, perhaps this issue could also apply to other types of chains, such as repetitive movements during strenuous exercise. It is at times such as these, when the exercise is particularly all-consuming and strongly sensory, that we know that naturally occurring morphine-like compounds (endorphins) are released, resulting in the well-known phenomenon of the 'jogger's high'. These naturally occurring opiates will have a net effect of inhibiting neuronal activity, and hence, at the neuronal level, reducing the size of an assembly.[49] But, at a more holistic level, vigorous exercise also has another effect on the brain.

It turns out that even brisk walking can drive the production of new neurons (neurogenesis),[50] whereby stem cells, the universal primary cells from which diverse cells originate, will convert themselves into neurons at maximum capacity, as well as stimulating the release of chemicals that help then-new cells grow. And it doesn't stop there. While vigorous physical activity increases the manufacture of brain-stem cells, additional stimulation from an enriched environment will increase the connectivity and the stability of those connections.[51]

We can now go one step further. If, as we've seen, the environment can change an individual's thinking, in the normal, healthy brain, then could the reverse occur and might the mental process of thinking actually change the physical brain itself? Remember how the 'mere' mental act of imagination of playing the piano had a measurable impact on the physical brain. A similar consequence of thinking, even in rats, can also be seen in an experiment in which Fred 'Rusty' Gage, professor in the laboratory of genetics at the Salk Institute, California, has proved that, in order for exercise to generate new brain cells (neurogenesis), exercise has to be voluntary.[52] The rat must proactively decide to engage with the exercise wheel and not just be forced to run on a treadmill against its will.[53] What might this literally incredible process be, whereby the mental can drive the physical?

Interestingly enough, it is only when rats are engaged in *voluntary* physical activity that important physiological factors come into play: most importantly, there will be an absence of stress, as suggested by the lower levels of anxiety they exhibit,[54] and, with less stress, there will be a reduction in all the related and potentially damaging hormones, such as cortisol.[55] Research has indeed found that regular voluntary exercise prevents stress-induced impairment in health and cognition in rats and mice.[56] Although we may still be unable to trace in any meaningful detail exactly how the absence of one set of chemicals leads to the all-important brain state for neurogenesis, nonetheless, a further, really fascinating clue is that it is accompanied by a certain brain-wave pattern (the characteristic EEG of theta rhythm),[57] which, in humans, is also indicative of a brain that is paying attention.[58]

So, if there is a link between plasticity and attention, and since attention cannot occur when you are completely unconscious, could there be a link between plasticity and consciousness? And, if so, could that link be the formation of appropriate assemblies? Plasticity of the brain is stimulated by an enriched environment, which, as we've seen, leads to greater branching of neurons, which, in turn, facilitates their connectivity. In our lab, we set out to investigate whether the dynamics of assembly formation could reflect such long-term plasticity changes over weeks in an animal previously exposed to an 'enriched' environment. And the answer was: yes.

When rats had been exposed to an interactive, enriched environment for three weeks,[59] they demonstrated a significant difference in behaviour from counterparts which had been placed in a far less exciting situation: a 'minimally enriched' world of a few polystyrene boxes. The rats who had had the benefit of interacting with ladders and wheels, and had been able to climb on little trapezes and branches, were measurably less stressed than those who had not had that opportunity: the maximally enriched and calmer rats, compared to the minimally stimulated ones, showed a significantly shorter waiting period before consuming novel food, and this change was accompanied by enhanced electrophysiological responses suggestive of large-scale activation[60] of neurons, that is, an enhanced assembly.[61]

Now, this observation raises an interesting question. If enrichment can enhance plasticity and perhaps assembly size, could the same occur the other way around? By having short-term, larger assemblies, could you make the brain more sensitive to plasticity? In other words, would it be easier to establish more long-term, localized, hard-wired connections? If that was the case, then that would help us to understand the survival value of consciousness, in evolutionary terms. After all, it is not immediately obvious why we can't just thrive in the world and navigate it as automata: why bother being conscious?

Let's invert the issue. This time, we won't think about consciousness and its survival value but just about assemblies. Assemblies, by encompassing many more networks of cells in the brain to work together in a large-scale collective, will lead to a widespread activation, which is one of their defining features: this, in turn, will enable much greater, more widespread adaptation in the brain, rather than just some little local response that would occur in a brain where assemblies (and hence consciousness) weren't operative. And, if that were the case, then that would mean the following: if assemblies are indeed correlates of consciousness, then, by being conscious, you are going to have a brain that is much more widely adaptive to the environment. So, the function of assemblies, and therefore the incidental tag-along correlate of subjective consciousness, would be to facilitate and coordinate a more widespread adaptation to the environment. And that would mean that animals that were capable of more – that is, 'deeper' – consciousness would be more adaptive. This makes intuitive sense: just look at us humans, with our planet-wide adaptability and, surely, the deepest consciousness of all species.

EXERCISE, PLEASURE, NEUROGENESIS – AND ASSEMBLIES

The next question is even more far-reaching than that of the evolutionary value of consciousness. Could the existence of neuronal assemblies in the brain help us understand the link between plasticity, neurogenesis, exercise and the all-important and most mysterious factor – the seeming necessity for a mental event, a conscious thought – to initiate these changes in the brain? In order to explore this question, consider these two possible scenarios, each of which relates to vigorous movement yet would involve very different-sized assemblies.

Scenario one is driven from the *outside*: the loud beat of the music, for example, or the sharp decline of the ski slope, or the heave of the surfing wave, which leads to high arousal, the physiological correlate of which, as we've seen already, is the release of dopamine. The high arousal caused by dopamine, the excitement of vigorous exercise in a fast-changing, sensory environment and the consequent release of endorphins (the naturally occurring morphine) will all conspire to reduce the size of assemblies, which is, in turn, correlated phenomenologically with an experience of pleasure in living in the here and now.[62] The greater and faster the literally 'sensational' stimulation from outside, and the smaller the assembly, the more the state of consciousness will be passive and un-self-conscious, simply reactive to whatever stimulus comes along in rapid succession – and the reaction in some cases will be bordering on fear. The assembly would be small.

In contrast, now reflect on an alternative, scenario two. This time, the drive to move is cognitive: it comes proactively from *inside*. You have had a specific thought – made a conscious decision to go jogging, for example. This voluntary decision, this active thinking, will, as we've just seen, provide the key requisite conditions for neurogenesis: the production of more neurons. The more neurons you have, the more potential for connections between them and the greater the connections, the more versatility for plasticity. We've also seen that the same stimulating, 'enriched' environment that drives plasticity will also drive greater assemblies and hence deepen consciousness. But the converse will also hold: a greater assembly will in turn facilitate the more ready formation of local synaptic plasticity, which will then last for a much longer time period than the extensive assembly itself.

If assemblies are indeed the neural correlate of consciousness – if you can't have one without the other – then we can see why conscious thought would be necessary for exercise-induced neurogenesis and, indeed, why a 'thought', such as thinking about playing the piano, can lead to the enhanced plasticity that was observed in experiments. Over time, the enhanced plasticity that ensues will give you the potential for greater-sized 'stones' and hence a deeper consciousness, in which the modulatory chemicals of arousal play less of a role, and the dialogue with the outside world, as in the piano-playing experiment, is weighted more on the inner brain processes than on external triggers in the environment. The assembly will be larger.

This proactive alternative – a deep thought and reflection induced by, say, jogging – will be in contrast to the mere 'mindless' reactivity of dancing with abandon: rather, it might provide a neuroscientific basis for the currently popular phenomenon of 'mindfulness'. Mindfulness is defined as 'a mental state achieved by focusing one's awareness on the present moment, while calmly acknowledging and accepting one's feelings, thoughts and bodily sensations'. Remember: at the beginning of your walk this morning, we saw that one of the cognitive effects of an urban environment is that it forces you to be more reactive to external distractions, while the rural environment offers the opportunity for you to be more proactive, to initiate your own 'voluntary' actions and thoughts.

Such was the case when you ventured outside: the gentle exercise of walking with Bobo allowed your inner mental processes free rein, without the distraction of a raised heart rate and fast-changing scenery requiring pre-emptive action. Ed Stourton's remark that dog-walking enables you 'to escape ordinary life' may well be all too true. But ordinary life is now, sadly, nudging you. You're going to be late for the office – and there's only time for only a very quick breakfast.

4
Breakfast

You can grab only a quick bowl of cereal and a hot coffee, to make the most of this narrow window of breakfast solitude. For the next twenty minutes or so as you eat, say, to the soothing sounds of Mozart, the powerful activation of your ears, eyes, tongue, fingertips and nose will be commandeering your consciousness. Of course, there are times that you are conscious without your senses being overtly stimulated, for example in meditation, or just concentrating in deep thought, but this is an occasional privilege for those able, in any event, to achieve a complete detachment from their surroundings. Most of the time, for most of us, consciousness is driven from one moment to the next by what is happening immediately around us, as the five senses buzz their reports ceaselessly into the ever-receptive, waiting brain. The senses more or less colour our every waking moment: they keep us in touch with the outside world and enable us to navigate it appropriately. If we return to the stone-into-the-puddle metaphor, the important issue we're going to focus on here is the *force* of the stone throw, how the senses and sensation, pure and simple, will shape consciousness. But straightaway we bump up against two baffling problems: one is to do with space, and the other with time . . .

THE FIVE SENSES: SPATIAL BRAIN FEATURES

The spatial problem is one of neuroanatomy: whereabouts in the brain the distinct senses will be differentially processed. This might sound simple enough: you either see something, hear it, feel it, taste it or smell it. We have five senses which are clearly recognizable as themselves and differentiated from the others. Yet even at this most basic level, in brain terms, the neuronal territory demarcated for processing the different senses is not inherently this specific or distinct. Even in human adults the various sensory systems can trespass across the official anatomical boundaries within the brain, between one and another: the visual cortex of blind people, for instance, is activated by the sense of touch, when they read Braille.[1] Moreover, it's well known that if you lose one sense, the others become stronger: the neuroscientist Helen Neville has demonstrated how deafness enhances sight,[2] and that deaf people use auditory brain areas to process touch and vision.[3] Meanwhile, blind people can discriminate pitch better than non-blind counterparts,[4] and they are more proficient at determining the location of a sound.[5] The visually impaired are also superior at other non-visual tasks, such as speech perception[6] and voice recognition.[7] And, in experiments on animals, deprivation of a particular sense has revealed that these enhancements can happen in a trice: rats will show improved hearing after just a few days confined in the dark.[8]

However, even when an individual is not lacking one of their senses, the brain can still play tricks with the processing of different modalities. Though very rare, the existence of synaesthesia (literally, a 'bringing together of the senses') has been recognized for several centuries: in this fascinating condition, a single input which the vast majority of people would perceive with only one sense will be experienced in two modalities. For example, colours and movement might be 'seen' on hearing sounds.

In brain terms, synaesthesia is clear proof that there must be an impressive flexibility in the way in which neuronal territory is colonized for processing the different senses. In this case, it's not that one area is invaded by another but rather that the connectivity between regions is unusually exuberant and excessive:[9] the result is that activation of one area – say, for recognizing letters – also causes direct activation of another, in this case, the one linked to colour recognition. Alternatively, it might be that there's some kind of pre-existing blocking mechanism between the different sectors of cortex that would normally ensure there was a clear segregation of feedback in order to prevent any ambiguity: but this otherwise impregnable barrier would break down in synaesthesia. If normal feedback were not prevented as usual, then any signals feeding back from late stages of multisensory processing might impact inappropriately on the earlier, single-modality stages: so now auditory tones, say, could activate vision. This disinhibition could also feature in clinical conditions such as temporal-lobe epilepsy, head trauma, stroke and brain tumours.[10]

In any event, the irrefutable reality of synaesthesia, along with the well-accepted over-compensation in those who are blind and deaf, leads to the inevitable, yet intriguing, paradox: while, *subjectively*, the experience of each sense is very different and completely distinct, the neuronal machinery *objectively* mediating these diverse experiences is standardized and interchangeable. Once an input from the outside world is converted, say, by the cochlea or the retina, into volleys of action potentials, these electrical blips are then pinged to different bits of the brain, where they end up in respective sectors of the cortex which are nonetheless modular and homogeneous in organization, suggesting a commonality in their structure and connectivity – a bit like a cookie cutter.[11]

So, where and when does the amazingly qualitative difference in subjective experiences intervene? Where and how does the subjective distinction of hearing something versus seeing it become possible? Somehow, a clear segregation in experience is made possible by the brain, even though the physiological processing mechanisms are pretty much the same for both senses. If we could resolve this paradox, it would help us to understand how the water turns to wine, how the objective relates to the subjective.

THE FIVE SENSES: TEMPORAL BRAIN FEATURES

The second puzzle is one of time: the senses will be processed in the brain at different speeds which you will nonetheless experience as simultaneous. You'll both hear and see a hand clap at what, subjectively, seems to be the same moment in time, even though auditory processing happens to be faster than visual: or, again, if you touch your toe and your nose at the same time, you'll consciously experience a single, multi-modal moment of consciousness, even though the signal from your nose reaches your brain first, since it has far less distance to travel. What this means is that we must have time windows (epochs) spanning each seemingly single moment of consciousness: an epoch will be a window of time over which the senses can catch up with each other to be combined into the familiar multisensory whole that we call a 'moment' of consciousness. Your brain will need somehow to synchronize things: there'll have to be an appropriate time lag to encompass all the different sensory modalities arriving at different rates and, inevitably, the slowest sensory signal will therefore set the pace.

It turns out that these epochs last for several hundreds of milliseconds. 'We are not conscious of the actual moment of the present. We are always a little late': almost half a century ago the ingenious physiologist Benjamin Libet made this observation when he was studying patients at a local hospital who had been admitted for neurosurgery and consequently had a hole drilled into their skull to expose the cortex.[12] In one experiment, Libet used an electrode inserted through the hole to stimulate the part of brain that would elicit the sensation of a tingle in various parts of the body. Amazingly, the subject didn't report being aware of the stimulus for up to 500msec, a half a second afterwards, and an eternity in brain time, given that an action potential will span only one thousandth of a second. Libet went on to show that when stimulation was applied to a remote part of the body such as the feet and the responses were recorded in the brain, then there was a marked time lag between the brain registering the event and the subject being conscious of it. And it's not just a question of having the brain register an incoming signal within a window that ensures that even the slowest-processed sense has finally arrived: awareness of consciousness seems to take even longer still. Research shows that when subjects categorize images presented in a random order as either animals or vehicles, although the brain has registered a difference early on, the 'conscious' decision occurs much later (peaking at around 250msec).[13] These epochs of consciousness of several hundred milliseconds at least provide just the right period of time for an assembly to form and then disband. An exploration of assemblies could therefore help us to understand what is happening in the brain.

Here is a series of images from our own lab which shows assemblies in different parts of the cortex at 300 milliseconds after their initiation in eleven different experiments: in the visual region, the highest final activity always ends up mainly in the deep layers, whereas, in the auditory system, the complete opposite occurs. The findings suggest a differential processing in hearing and vision that is *not* readily reducible to conventional synaptic events but instead to an emergent property of larger-scale collective neuronal dynamics that would have passed undetected were it not for the optical imaging of assemblies.

But what makes the *quantitative* 250–300msec period of time which must apparently pass before consciousness usually registers so *qualitatively* different in being linked to a subjective experience? If they are indeed necessary for consciousness, then the very nature of assemblies themselves might hold a clue. Neurons within assemblies are not operating like isolated, long-distance telephone cables, independently relaying a single message as soon as it arrives at one neuron to far-flung destinations in the brain. Instead, an assembly is self-organized and intrinsically self-sufficient over the hundreds of millisecond it takes to grow. The mounting activity will be passed on only locally, which is why it slowly spreads, like ripples. The

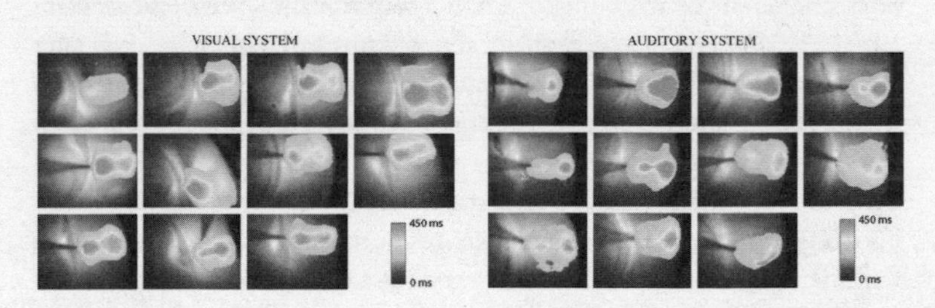

Fig. 8: Optical images showing a difference in assemblies between the visual and auditory systems in eleven different experiments.[14] The frames were taken at the time at which the fluorescence signal of a particular location finally decreased to beneath 20 per cent of the maximum value: 300msec. Note that this activity is concentrated in the deeper layers in the visual cortex and in the more superficial layers in the auditory cortex. (To see these scans in colour, see Plate 6.)

important point is that the subsequently emergent ripples (the degree of consciousness) will also be summed over a window of time, an epoch: no wonder a moment of consciousness takes up to half a second.[15]

But then we still have the problem of space – of how the *objective* cookie-cutter arrangement of the cortex can correspond to the *subjective distinctions* of hearing and vision. But what if the distinctive perceptual experience characteristic of either sense was in some way related to differences in the properties of neuronal assemblies in the visual and auditory cortices which are now apparent only after several hundred milliseconds: if this were the case, we could now differentiate the phenomenology of hearing and vision using the *same* yardstick of objective physiology. But then, where or what is the single yardstick for the corresponding subjective sensations of hearing and seeing? If you had to describe to a Martian the subjective difference between the experience of hearing and seeing using some kind of common unit or frame of reference, what would you say?

Until we can come up with an answer, it is very hard to map the phenomenology on to what we are monitoring objectively in the brain. Still, I have one suggestion: in physiological terms, vision is primarily (but not exclusively) discrepancies in detection of spatial edges, whereas hearing is primarily (but not exclusively) detection of time differences. If so, then the spatial features of assemblies, varying as they do so crucially over a particular time window, could help us devise a new addition to the neuroscience toolkit. The eventual ideal would be to work out a single yardstick of space-time, some kind of descriptive mathematical equation that can also be applied to an epoch of subjective consciousness.

MULTISENSORY CONSCIOUSNESS

But then, what would that consciousness actually *be*: a single sense, or all five combined? Everyone would agree there are five different types of sensation, so a reasonable deduction might be that consciousness must work that way too, that the brain must maintain some kind of parallel processing system, forever segregated into five separate senses, which then translates into respective and distinct streams of subjective experience. This seemingly straightforward reasoning, as we've already seen, was presumed by the late Francis Crick and his colleague Christoph Koch, who strove to identify a neural correlate of consciousness specifically for 'visual consciousness', an apparently separate phenomenon that could operate independent of the other senses.[16]

Even though it is almost impossible to imagine real-life scenarios in which we would use only one sense, the classroom has seen just that assumption – that the senses work independently – put into operation. Back in 1978, a new approach to learning based on this very premise was introduced. The idea was that the brain could be compartmentalized into three 'learning styles': visual ('V'), auditory ('A') and kinaesthetic ('K') – 'VAK'. VAK was originally proposed by American educationalists Rita and Kenneth Dunn over thirty years ago as a way of both explaining individual differences in learning abilities in children and promoting a new way to cater for them.[17] Teachers reported that their pupils inevitably changed the emphasis they placed on their senses depending on the subject being taught. In art, for example, a premium is placed on visual attention, while it would be acceptable in many instances for students to shut their eyes in a music lesson. However, the Dunns' theory went much further, with the idea that some individuals are by their very nature primarily 'visual' learners, some primarily 'auditory' and others primarily 'kinaesthetic', meaning they favour using their sense of touch.[18]

No independent research has found that any learning improvements have resulted from the VAK initiative, and the only real advantage seems to be that of teacher enthusiasm. So why should anyone find the pedagogic distinctions of the VAK paradigm appealing or, at least, convincing? The rationale stems once again from the deceptive notion of autonomous brain structures or, in this case, what could more loosely be termed 'modules' of brain regions all having independent functions. It is certainly the case that, over millions of years, brains have evolved a number of seemingly specialist structures: modern humans have appropriated many of those structures and attributes to do the things that only we humans do. What defeats the whole VAK rationale, however, is that these functional modules work only by virtue of their interconnectivity: they do not work in isolation.

A good illustration of what turns out to be a much more complex and interactive process than the VAK proponents suggest was demonstrated by the cognitive neuroscientist Staneslas Dehaene.[19] He asked his subjects to repeat an arithmetic calculation during brain scanning, an exercise often used by anaesthetists as their patients are succumbing to unconsciousness, which is to take seven away from a hundred and then to continue the process of subtraction from the respective number left. As you lie there, you are told to say to yourself, 'One hundred take away seven; ninety-three take away seven; eighty-six take away seven,' and so on. Those undergoing surgery rarely get to the end, but most of us testing our mental agility while fully awake would try to see how fast we can rattle through these 'serial sevens', which seems like a simple process. Yet when Dehaene used brain imaging to look at the regions of significant cerebral activation when even that one-step calculation was performed, it turned out that as many as ten to twelve different areas of the brain lit up. In other words, it was yet another study to reveal that we use our brains holistically.

Interconnectivity includes a vigorous interplay between the connections streaming into the brain from our sense organs: if we look at ourselves as primates in VAK terms, our brain processes are strongly visual, taking up approximately 30 per cent of the cortex, compared with 8 per cent devoted to touch and 3 per cent to hearing.[20] What we do with incoming visual information is to construct spatial 'maps' of the world so that we understand how things connect together. That is true even for people who are congenitally blind: they construct these type of maps, too. Obviously, the blind do not get their initial information visually but by touch and hearing – yet they treat this data in the same way within their brains as sighted people. The blind will construct maps of the world to understand where things are located both actually and conceptually.[21] So there is a multisensory, 'cross-modal' process in which information, whether it arrives kinaesthetically, aurally or visually, is interrelated and becomes one information pack. This cross-modal processing is an established and basic feature of brain function.

All of us are familiar with detecting when things are in sync and when they are not, such as when the lip movements of a TV reporter are out of sync with the sound. In one experiment, activity in the brains of subjects looking at and listening to either in-sync or out-of-sync lip movements was compared.[22] Both the visual and the auditory cortex have areas that, in scans, show an additive effect when the sensory input is in synchrony or a subtractive effect when they are not. This talent for spotting a discrepancy makes a lot of sense in evolutionary terms since visual and auditory information need to be closely linked. In the dark, you need to be able to locate the source of a sudden sound; after all, the unexpected noise could be your next meal – or the unfortunate reverse.

Moreover, being able to read lips helps you to hear despite there being a large degree of background noise.[23] The actual brain process underlying this effect is revealed in recordings from awake monkeys who were watching or hearing naturalistic audiovisual stimuli: the researchers found that the patterns of responses from individual brain cells became more reliable when multisensory stimuli were presented. In contrast, when the visual stimulus did not match the sound, this multisensory enhancement was much reduced. So, multisensory influences enhance information processing even in parts of the cortex (the primary cortex) which, traditionally, have the basic job of the initial processing of a single, sensory modality.[24]

The late educational psychologist, John Geake, carried out a simple experiment to prove this sensory interconnectivity.[25] He instructed a group of children to look at two groups of objects; in both, there were too many items for them to count with any precision in the time allocated. Even so, when subjects were asked to say which group had more objects in it, even participants as young as five years could give a fairly accurate answer. The key discovery, however, was when one of the groups of objects was replaced with a series of rapid sounds which, once again, were too quick to count. Interestingly enough, in this case, when the comparison was between a visual and an auditory input, there was no change in the children's accuracy: they could still determine which group had more. The clear conclusion was that the specific sensory route through which the information has arrived is irrelevant. The brain will process the inputs at 'higher' levels of brain operations – and 'abstract' it, so that it is the message content, not the medium, that is all-important.

While we can be aware of the five different senses, our brains, nonetheless, will *not* usually need to dissect and segregate them during a conscious experience. Learning – all types of thinking, really and therefore consciousness itself – involves a process of abstraction: irrespective of the sensory mode by which we receive the information, we extract the essential meaning to decide what action to take before committing to some kind of response. A good example is the history we all learn at school. Most of us know something about our country's constitution and milestone events in its past, but whether we acquired that knowledge by reading books, listening to teachers in lessons, or perhaps by watching a documentary film, it will have mostly faded from memory because it is, in fact, irrelevant. Another example of this 'abstraction' might be when you are walking in the woods on a spring morning, smelling the crisp air, seeing the bright green leaves, and so on, and consequently feel a general sense of well-being. You do not feel any need to distinguish between the various senses involved. A moment of consciousness is multisensory: the whole is more than the sum of its five different inputs.

If anything, when it comes to that moment of subjective consciousness, the different senses enhance each other. This synergy of senses is particularly obvious when, at the current moment, you are eating breakfast and drinking coffee. As anyone who has had a heavy cold will attest, a blocked nose renders food tasteless. The sensors in your tongue might be relaying their messages well enough but, if your nose is out of action, then you cannot really experience any flavour at all. There's another simple way of demonstrating that the nose and tongue normally work together: if you keep a jelly bean, say, in your mouth and simultaneously hold your nose, you'll experience sweetness but no flavour. However, once you free up your nasal passages, the flavour will flood into your consciousness.[26] Flavour may be more or less important, depending on the identity of different substances: for example, coffee, chocolate and garlic cannot be recognized as such when the nose is blocked, while whiskey, wine and vinegar can.[27] Moreover, different odours can change the final taste experience.[28]

The nose is of key importance because chewing releases energetic molecules that travel through the back of the mouth to receptors in the lining of the nasal passages. The range of possible smells is much more diverse than the options on offer for taste. The human tongue has only five different receptors: sweet, sour, salty, bitter and *umami* – a Japanese term meaning 'delicious taste', coined in 1908 by chemist Kikunae Ikeda, who worked out that the all-important component of *umami* was glutamate, and then patented the notorious flavour enhancer monosodium glutamate (better known as MSG). This limited repertoire of options is nothing compared to the seemingly endless possibilities within the nose, where there are hundreds of receptors for odours. This means that there is an almost infinite range of possibilities for mixing and matching taste and odour, leading to the wide variety of sensations we subjectively describe as flavour.

However, as we all know, while smell and taste might well be experienced synergistically as flavour, they can also be perceived independently as taste and smell. Smelling through the nose and smelling through the back of the throat while tasting are neurologically different phenomena. If the smell and taste signals arrive in the brain simultaneously, then odour and taste are integrated into the experience of a flavour. No merging of these two senses will occur when you simply sniff food that's not in your mouth, or when you see food without tasting it. Yet even when you are not able to smell what you're about to ingest, vision alone will nonetheless also have a huge impact on taste.

Recent research shows an intriguing synergy between vision and taste: it turns out that the mere colour of a cup can influence the taste of hot chocolate: the drink tastes more chocolatey in an orange or cream-coloured mug compared to a white or red one. Amazingly, our senses perceive food in a different way depending on the characteristics of the container from which we eat and drink it.[29] Even the simple, physical shape of an item of food can make a difference: for instance, if a piece of cheese has pointed rather than rounded corners, it will taste sharper.[30] Less colour contrast makes things taste sweeter, for reasons that are still unclear: hence white yogurt on a white spoon, say, tastes sweeter than pink yogurt on a white spoon.[31]

And if vision can play such an important part in our other senses, then perhaps it won't be so surprising that hearing also has a significant impact on the experience of tasting something. For example, a quiet ambience can enhance saltiness, while loud background noise diminishes it; a loud background will decrease the taste of sweetness but enhance the intensity of perceived crunchiness. And as many people have probably already discovered, the sound of coffee grinding or the crunching of crisps (congruent sounds) enhance the taste of coffee and crisps, while incongruent sounds will have the opposite effect. This phenomenon even extends into more general cultural contexts: one study found that when French wine is sold to a background accompaniment of accordion music, over three times as many bottles were sold compared to when German Bierkeller music is played.[32]

The sense of touch can also escalate the experience of overall subjective taste. After all, we experience and differentiate foods through our physical contact with them. In particular, fats give that creamy 'full-mouth feel' of butter and ice-cream, as will the oiliness and viscosity of a rich salad dressing: accordingly, an area of the brain contains neurons that respond specifically to the texture of fat in the mouth. These brain cells are activated not only by fatty oils in the mouth, such as vegetable oil, and foods rich in fat, such as ice-cream and chocolate, but also by non-food substances that have a similarly oily texture. The taste of a fizzy drink is also greatly influenced by the sense of touch, albeit in a different way: a flat beer will taste very different from one with carbon-dioxide bubbles still rising in it. More generally still, the froth, viscosity and smoothness of different foods and drinks will contribute to the experience of eating them.[33] Even the touch of different utensils used for eating is important: for example, yogurt eaten from a lightweight, plastic spoon tastes more filling and seems more dense than from a silver spoon.[34]

A close partner to the sense of touch, and a further conspirator in the holistic consciousness state of eating or drinking, is temperature. Very cold ice-cream has little flavour, for example; warming it increases the perceived sweetness. The molecular processes in the taste buds of the tongue which play a key role in the perception of sweet, bitter and *umami* also regulate sensitivity to temperature. Even raising the temperature of food to between 15°C and 35°C enhances the neural response to sweetness. In about half of the population, heating or cooling the tongue is, by itself, enough to trigger different taste sensations: warming the tongue results in a sweet sensation, and lowering its temperature induces a sour or salty taste.[35] Moreover, touch and temperature work closely together. The allocation of brain territory within the section of 'somatosensory' cortex which relates to touch is greatest for those very same parts of the body that would be involved in sipping a hot drink: the lips and the hands. So imagine the amount of stimulation flooding into the brain when you sip from a hot mug of chocolate that you are cradling in cupped hands: an additional force will now be added to the throw of the stone.

One interesting suggestion has been that the five different senses involve different amounts of 'consciousness':[36] vision scores highest, followed by taste, touch, sound and, finally, smell. But the way the term 'consciousness' is used here is quite misleading. Consciousness involves not just the degree of direct sensory experience but also the contribution of personal significance. As the anthropologist Clifford Geertz summed up so beautifully, 'Man is an animal suspended in webs of significance he himself has spun.'[37] Much as we can speak of the particular factor of throwing the stone with varying force, even when it comes to the raw senses the *size* of the stone (the cognitive and contextual associations of a stimulus) will be highly significant. So we can revisit the ranking of the senses, but not so much according to amounts of 'consciousness', rather according to how much each relies on a specific context and meaning.

Take vision, which is surely the most specific and least abstract of the senses. The world around us is composed of edges, patterns, shades and brightness, and these coloured shapes will usually have a 'meaning', be a recognizable object or person. And what you see, as we discussed in the previous chapter, will invariably 'mean' something to you that is personal to you alone. Unless you are looking at a plain, homogeneous panel of colour – which is in itself very rarely the case – then there will always be a context: a highly specific, unique and unambiguous scene. When you look at a table, you're looking at a particular table in a particular place at a particular time. When you look around you, you're not just seeing generic colours and shapes that would be just the same somewhere else at some other time, rather, you are accessing your personalized connections for a particular moment in your life: the stone, when it is thrown, will be relatively large.

Next is taste: since this is such a basic sense, it might be surprising that it ranks as the second most personalized. But then again, the context is highly defined: you are tasting a very specific item of food or drink. One factor in determining taste is the timing of one experience in relation to another. In one study, subjects evaluated a sample of lemonade in terms of whether it was sweet or sour. Then they were asked to drink a further sample of lemonade which contained additional citric acid. When they were given a third drink, actually the original balanced sample of lemonade, they judged it as being by far the sweetest of the three.[38]

As we've seen, taste can be much influenced by the circumstances under which the tasting takes place, and in real life this will always be massively shaped not just by smell but by the other senses, too: the colour of the cup, the sound of accordion music in the background, the froth and creaminess in your mouth. And because taste is the product of, and constrained by, the other senses, then those senses in combination will invariably define a specific context, and one, therefore, that is dependent on associations: once again, a larger stone.

Sight and taste have been defined respectively as being roughly 90 per cent and 80 per cent 'conscious', but a more accurate term might be 'context-dependent'. The actual percentages are meaningless: it's their relative positioning compared to other senses that counts. Touch ranks much lower, for example, at less than 50 per cent.[39] Yet this is not surprising if we once again substitute the term 'conscious' with 'context'. The touch of velvet, silk, wood bark or bare skin can be experienced in numerous different situations. While you may be able to be 'conscious' of the object you are touching, the rest of the context in which that object happens to be, is irrelevant. More emphasis now is on the direct sensation emanating from the interaction of your body – most likely, your sensitive fingertips – with the surface in question: the stone can be smaller, with the importance now placed on how much more forcefully it is thrown.

Hearing comes in next, at 30 per cent: bearing in mind that we must not put too much emphasis on this percentage itself, it is valuable to compare it with the other senses, in that hearing ranks fourth and hence lacks a predominantly contextual component. Compared to vision, taste and touch, hearing is, after all, passive. You can open and shut your eyes at will, turn your head proactively, put food in your mouth and chew it or stretch out a hand to feel something: but you can't turn your hearing on and off. In all cases, the sound finds you, not the other way around. Less networking is required, and the stone is ever smaller. So if hearing is one of the most 'passive' senses, it would follow that it is easier to experience in a state in which anaesthetics have reduced the formation and size of assemblies: it is indeed the final sense to cease functioning under general anaesthesia and also the first to recover its function as the patient is awakening.[40] Perhaps this is because hearing, the fastest-processed sense,[41] is able to act on its own as the stone thrown into the puddle: the extent of the *force* of the throw is paramount.

Finally, there is smell, which, at only some 15 per cent 'conscious', is the most context-free of all the senses. Interestingly enough, loss of the sense of smell is one of the earliest signs of Alzheimer's disease,[42] because the pathway connecting the nose and the brain goes directly to the 'limbic system'. The limbic system is an extensive, interconnected cluster of brain structures which link with the early stages of memory processing and, most significantly here, emotions. It is hardly surprising, then, that smell can elicit such powerful and immediate feelings and, as such, is the most primitive of the senses.

The contribution of smell seems to have taken something of a back seat in the sensory repertoire of our species, compared to other animals. We humans have, for example, far fewer olfactory receptor cells and olfactory-related genes than dogs and rats. Smell is without doubt a powerful, primeval sense which allows an animal instantly to determine whether something is rotten, wounded or sexually receptive, and to track prey over longer distances. No wonder that other senses have taken over in humans, in whom instant, atavistic reactions are tempered by knowledge and thought. However, some have argued that, compared to other species, humans may after all have larger brain areas associated with olfactory *perception*.[43] So the amount of olfactory processing in which we still engage might be more frequently applied not so much to the processing of raw stimuli with the kind of power and precision we see in dogs, but to memory processes for which we have the follow-on cerebral machinery way greater than any other species. Then again, even in us humans, the 'subconscious' emotional effects of smell should never be underestimated.

Take pheromones. These sneaky chemicals are released in all species in a variety of contexts ranging from survival by means of territorial marking to alarm signalling to reproduction, at least in non-human animals. For humans, pheromones mostly come into play in the preamble to social and sexual behaviour. Although the mechanisms by which pheromones function remain controversial, there is evidence that these chemicals do indeed affect us in surprisingly precise ways.[44] For example, using just the sense of smell, human subjects can detect individuals that are blood-related kin. Participants in a study who had never met the individuals from whom odours were being tested were able to match correctly mothers with their offspring but could not reliably match the odours of husbands and wives.[45] Mothers can identify by body odour their biological children but not their stepchildren. Preadolescent children can detect by smell their full siblings but not their half-siblings or step-siblings, a phenomenon that perhaps explains how incest is avoided.[46] Babies can recognize their mothers by smell and, in return, mothers and other relatives can identify a baby by their odour.[47] This primitive recognition between blood relations is not due to cognitive factors – after all, a baby can't have a developed memory – but surely reflects a more primitive bonding. In this case, the stone is small, depending for its effect not on personalized links from prior experience but on the sheer force of the throw: the raw activation itself of the sense of smell.

So what does all of this tell us about consciousness? The activation of the different senses will equate to a different 'force' of stone throw, in combination with a different 'size' stone. Vision, by virtue of being highly context-dependent, would be a large stone, not necessarily strongly thrown, while smell would be the opposite extreme – a strong raw sensation without immediate and obvious context: a smaller stone. But a small stone thrown with great force can still generate extensive ripples: perhaps one of the best examples of this is music . . .

MUSIC AND THE BRAIN

Music is defined by one dictionary as 'vocal or instrumental sounds (or both) combined in such a way as to produce beauty of form, harmony, and expression of emotion'. Yet this definition doesn't really help us understand the colossal significance of music to our species, a significance so huge that the music industry is in fact economically bigger than the pharmaceutical one.[48] In the age-old debates about what makes us human, some argue that sign-language skills are possible to a limited extent in other, specifically trained primates, but no one has yet claimed that chimps, or any other of our near-cousins, separated from us by only a per cent or so difference in DNA, are capable of generating and appreciating music. In contrast, our own ancestors have been doing so for tens, if not hundreds, of thousands of years.

A bone flute discovered in Slovenia has been estimated to be between 43,000 and 82,000 years old.[49] While the flute is quite a complicated instrument, more basic, and common, musical instruments in today's hunter-gather societies are rattles, shakers and drums.[50] So, presumably, these instruments would have pre-dated the flute, and musical activity would have been already taking place for a long time before this. Palaeontologists believe that singing would have existed before any type of instrument-construction by half of the intervening time: if so, then music-making could go back 150,000 years or more, beyond the estimated 100,000 years over which we *Homo sapiens* have walked the planet. Interestingly enough, the Slovenian bone flute was found on a Neanderthal burial site so, quite possibly, musical activity first appeared anything up to a quarter of a million years ago, starting not with our particular species but the entire genus *Homo*.

Music is clearly intrinsic to our lives. But does that mean it has true evolutionary value as a basic part of our make-up or is it instead an evolutionary by-product',[51] 'auditory cheesecake',[52] as the psychologist Steven Pinker dismissively suggested: agreeable enough, but hardly an evolutionary essential? While music is irrefutably a pleasurable activity, this could simply mean that it has hijacked the reward systems of the brain, just as recreational drugs have done, thereby tapping into pre-existing 'pleasure channels' which had originally evolved for other purposes of survival, such as eating and sex.

There are two different arguments which support the idea that the very basis of music is neither biological nor of any evolutionary value. The first is the notion that music could vanish from our species and the rest of the human lifestyle would remain virtually unchanged.[53] The second stresses that music is so specific to particular times and places that it surely has to be an artefact of each respective culture and is therefore not 'natural': there seems to be no guarantee that all the forms of music that have ever been contain identifiable common properties that would have been invariant since music came into being.[54]

On the other hand, it is hard to explain why, in the words of David Huron, professor of music at Ohio State University, 'There is no human culture known that did not, or does not, engage in recognizably musical activities.'[55] So, if we're to understand how music impacts on consciousness, whether it is a generic feature in our mental processing, we need to solve the riddle of why music is indeed a 'universal behaviour'[56] yet one that, firstly, has no obvious survival value and, secondly, is so wildly diverse in its different cultural manifestations.

Let's start with the evolutionary pointless 'cheesecake' possibility. Contrary to this line of thought, there are various reasons why music might, after all, have helped us survive as a species: Darwin himself suggested that music may have arisen due to sexual selection in mating calls. And, as well as mate selection, we can include social cohesion, group effort, perceptual development and motor-skill development.[57] Robin Dunbar, an anthropologist at Oxford, places music and dance in the same category as religion and storytelling: they are activities which encourage 'social cohesion'. Together, these behaviours improve our ability to have self-awareness and understanding and to form social groups which protect us from potential predators.[58] Yet, although social skills have been learnt in other species without the need for music, one response to this apparent anomaly would be that such social cohesion hasn't occurred to the extent and sophistication that it has in humans: we can cope with greater complexity of social interaction and, most significantly, this talent is aided and abetted by music. Perhaps, then, the value of music comes from exclusively human activities such as campfire singing and dancing. These communal experiences enhance cooperative survival strategies[59] and communication across the generations that does not occur to a comparable extent in other species. If these activities are indeed vital to the survival of *Homo sapiens* and if they can, after all, capture some biological 'universal' common property, then we can start to understand the brain mechanisms underlying these 'universals' in order to gain insight into how music can eventually impact on consciousness.

The second objection to music as being integral to our human nature is that it is too culturally diverse to be such a basic feature. However Ian Cross, a musicologist at the University of Cambridge, claims that there *is* a common factor in all types of music, namely, 'a level of temporal organization that is regular and periodic'.[60] If this is the case, then it is hardly surprising that instruments that enforce basic rhythm, such as rattles, shakers and drums, were among the first type of instruments to appear. But why should engaging in an experience of repeated rhythm be so essential and beneficial? If we could answer this question, it would help us understand how essential music is or isn't to the human brain, and whether it should or shouldn't be considered as inducing a particular, different type of consciousness.

One suggestion is that, in early brain development, periodic timing mechanisms are needed to establish a 'hierarchy of motor rhythms'.[61] The value of rhythm in any caregiver–infant interaction will then be that the infant follows and responds in kind to temporal cycles in vocalizations and movement from the carer that wouldn't be available in normal speech. It is the most natural sight all over the world to see a mother bouncing her infant up and down on her knee, while perhaps holding on to the little hands, moving them up and down to the rhythm of some nonsense nursery rhyme that is, at the very least, half sung or chanted. Moreover, the repetition of these actions, impossible in normal language, would enforce and engage the all-important plasticity of specific neuronal connectivity in the newly minted brain. And, along with this basic rehearsal in sensory-motor coordination, would come experience in interpersonal interaction and communication skills. Cross sums up music as 'a natural drive in human social-cultural learning which begins in infancy'.[62]

Everyone seems to agree that the experience of music will inevitably entail movement. If, as we saw in Chapter 3, 'Thinking is movement confined to the brain,' then music drives that movement unleashed from the brain into our bodies. Nietzsche claimed, 'We listen to music with our muscles,' and it is hard to imagine not at least tapping one's toe or swaying slightly to the intoxicating rhythm of music, if not standing up and, sooner or later, dancing. At its most basic then, exposure to, and experience of, music enables the young brain to develop in a way that could not be duplicated were music not to exist at all.

So if it is so essential, what mark does music leave on the brain, and hence the mind? One way to answer this question is by looking at the effect of different cases of brain damage on music appreciation. For example, one patient lost his emotional response to his favourite music following damage to a particular brain region (in his case, the left amygdala).[63] Similarly, patients with damage to the amygdala showed less 'fear' when listening to a piece of music that would normally have evoked a strong, negative reaction.[64] In another study, using imaging, subjects were asked to listen to music which they knew gave them 'the chills' – despite the weird term, an apparently pleasurable frisson evoked by certain musical passages.[65] Although it may not sound very scientific, this subjective sensation is accompanied by objectively measurable changes in heart rate, respiration and other measures of arousal.[66] On the other hand, in a study such as this, where participants were allowed to select their own music inducing strong emotion, any catering for subjectivity will inevitably not be standardized regarding individual associations and memories. Suffice it to say that, as these subjective 'chills' increase, many brain regions are activated (the amygdala and prefrontal cortex, among others), which are in turn associated with reward, motivation, emotion and arousal, as well as other pleasurable outcomes.[67]

Taken together, these findings show that music may mimic

biologically rewarding stimuli in its ability to engage similar neural circuitry in an area such as the ventral striatum which is well-known to be related to pleasurable experiences[68] like eating chocolate[69] or taking drugs such as cocaine.[70] However, the decreases seen in the activity of the amygdala, for example, suggest that the positive feeling might also be due to the blocking of basic fear responses. Interestingly enough, music has been identified as one of the only examples of a positively arousing stimulus that can decrease activity in this brain region.[71] So, the 'pleasure' of music may be due both to a push and a pull: on the one hand, a pull to positive activation of brain areas and circuitry related to reward and, on the other, and simultaneously, a push in that it suppresses networks linked to fear and other negative emotions.

It is hardly surprising that so many areas of the brain are activated during such an emotionally charged activity. No one would really expect to discover an *exclusive* area of the brain that is selectively and solely lit up by music. So perhaps the secret of music processing in the brain is the particular *combination* of brain structures it recruits, coupled with the suppression of areas relating to fear. By virtue of its 'universal' beat, tonality or harmony, music will set in train a repeating and reliable cycle of anticipation and pleasure. To the long list of participating brain areas we could also now add the cerebellum, the cauliflower-like structure at the back of the brain. This seemingly semi-detached mini-brain is a conspicuous feature of all vertebrate brains and has been dubbed the 'autopilot', since it is linked to sensory-motor coordination of the most automated type. If you're 'unconsciously' tapping your toe to the rhythm of the music, most likely it will be your cerebellum that is enabling you to do so.

Related to this effect is the fascinating discovery that music can help temporarily, yet very powerfully, alleviate the movement malfunction manifested in Parkinson's disease.[72] Why might this be? The area degenerating in this cruel condition is a central part of the brain related to internally initiated 'voluntary' movement, but the cerebellum is spared these neuronal ravages. Accordingly, it's well-known that if you ask a Parkinson's patient to perform movements driven by external stimuli – say, to place their feet on markings or sheets of paper over the floor – then, almost miraculously, they seem to be able to walk normally.[73] Music could be fulfilling the same role as these pieces of paper,[74] albeit working through hearing rather than vision: assisted by a continuous stream of external auditory stimulation, the patient's movements will lock in and respond accordingly. Another, not mutually exclusive, possibility is that the pleasure caused by the music might be enhancing the dwindling supplies of the key chemical that is deficient in Parkinson's disease: dopamine.

Needless to say, if music triggers emotion, particularly pleasure, it would be surprising if this particular hard-working and ever-popular transmitter were not involved. Dr Valorie Salimpoor and her team at the Rotman Institute, Toronto, reasoned that if music can arouse feelings of euphoria and craving similar to tangible rewards that involve the dopaminergic system, then listening to music should release brain dopamine. Her team subsequently found that endogenous dopamine is released (in the striatum) at the height of emotional arousal when a subject is listening to music. They also discovered that intense pleasure in response to music involves the *anticipation* of a reward, which results in dopamine release in an anatomical pathway that is distinct from that associated with the peak pleasure hit itself.[75]

As long ago as 1956, the author, composer and philosopher Leonard Meyer suggested that this 'tension' – the anticipatory build-up followed by a positive resolution – was at the root of emotional experience and, extending this idea, that the movement from a dissonant chord to resolution could account for the pleasure in classical music.[76] Music will provide a non-threatening sense of anticipation with each cadence, followed by an expected and repeating response. In the previous chapter, I suggested that pleasure and fear were very close bedfellows and that the crucial difference in determining whether an experience resulted in pleasure or fear was the degree of predictability and similarity in the genre of successive stimuli. Music, more than any other stimulus deemed pleasurable, would meet these requirements.

But then, while it is easy to see why the rhythms of eating and the successive taste of food might be pleasurable, the positive effect of merely repeating sequences of sounds is less obvious: what is it specifically about music that makes it so much an exclusively human activity, and why does our species spend so much time listening, playing and dancing to it? I would venture that music is directly comparable to language and that the uniquely human pleasure – and, indeed, survival value – of music can be understood best as an equal yet opposite counterpart to the spoken word.

Activity in the right hemisphere has long been correlated with emotion, and the fact that this side of the brain is sensitive to music has led many musical theorists, philosophers and neuroscientists to link emotion with tonality. This suggestion is reasonable enough, since the tones in music can be viewed as a characterization of the tones in human speech, which in turn indicate emotional content, prosody: musical tones might simply be exaggerations of normal verbal tonality.[77] Further similarities between music and language are still more obvious: both activities are essentially exclusive to our species, and both are conspicuous more by their incredible diversity across different cultures, historical eras, and so on, than by any overtly common features. Both have clear, culturally dependent rules and frameworks for their expression such that they need to be acquired by the young human brain after birth, as they are both created in a certain society and age. But if music is so closely comparable to spoken language, why have it as well?

There are crucial differences between these two essential forms of human communication which suggest that they complement rather than duplicate each other. While spoken language initially evolved to enable effective interaction with a small number of people, music is a much wider, collective means of signalling. And, while spoken conversations are unpredictable and completely unique, music can be repeated and, as we've seen, has reassuring cycles of anticipation and resolution. Most significantly, however, music is not restricted, as language is, to describing highly specific facts or ideas. Furthermore, music can, more probably, generate emotion without having to call up specific memories: we've seen already that hearing ranks rather low in being context-dependent. As the late neurologist Oliver Sacks so eloquently put it, 'Music has no concepts, and makes no propositions; it lacks images, symbols, the stuff of language. It has no power of representation. It has no relation to the world.'[78] David Huron concludes even more succinctly: 'Music can never attain the unambiguous referentiality of language, nor language the absolute ambiguity of music.'

In this way, language and music are two sides to the unique coin of expression and communication that is the birthright of our species, and utterly complementary in their equally important, yet different, roles. Music amplifies, exemplifies or reinforces the course of an ongoing experience. It provides a way of being in the here and now, unlike language, which is invaluable for referring to things you cannot detect directly with the senses. But, on the other hand, unlike the immediacy of white-water rafting, music is non-reactive: there are no unexpected stimuli for which you must take immediate action. If you respond at all, it is, if anything, not retaliatory but reflects or amplifies what you are hearing, as you hum along, tap your feet, sway your body.

If music is in the here and now and yet different from the interactivity of, say, downhill skiing, then to what extent will this ongoing experience be 'sensory' or 'cognitive'? In other words, how deep would be the ensuing consciousness – how wide the ripples in the pond? Surely the answer would depend on the type of music in question, which would in turn depend on different societies and generations having different reactions. For example, classical music that lacks verbal accompaniment has a less defined context and emphasizes the purely auditory, a sense that we have seen already is in itself one that feeds into the emotional and the subconscious, freed up from specific or literal scenarios: a forceful stone, but not necessarily a large one. The stone would be very small but forcefully thrown and followed by further small stones thrown in rapid succession as the beat and rhythms recur insistently, perhaps eliciting idiosyncratic and flimsy associations.

But you are heedless of all these neuronal machinations. There you are, driving along as Mozart washes into your ears and swirls on into your brain, conjuring up half-thoughts, sensations in joyful, illogical succession, while your eyes, hand and feet navigate you through the traffic effectively on their own. Yet, suddenly, something up ahead intrudes on your consciousness; your precious inner world, cosseted and curated by the music, now dissolves . . . at the sight of your office block.

5
At the Office

The time for music and the savouring of sensory pleasure in your private world is over: now you need to turn and face outwards, to embrace an external environment, a monolithic glass building, conforming to the style of so many office blocks built in the middle years of the last century. As you enter the edifice, your spirits plummet in free fall: how come this concrete-and-glass space looming around you is making you feel so miserable? You make your way to your desk: a work station identical to those stretching out either side in a long row in the open-plan office – where you will sit hunched in front of a screen for the seeming eternity of the rest of the day.

We spend, on average, over eight hours a day at work,[1] so the workplace will be an appropriate setting in which to examine the impact of the environment on our specifically human mindset. Probably, no one would be surprised that pupils in classrooms with natural light perform better than those being taught in artificial illumination, or that the design of hospital rooms might influence how quickly patients recover. But, despite having these intuitive insights, we do not really understand why this is and what is actually happening in the brain to account for these effects. Although it might seem common sense to encourage a cross-fertilization of neuroscience with architecture, the few studies that have attempted to study the relationship between the two have not been adequately rigorous.[2]

Yet surely this is a vital issue. The sheer brightness, loudness, and so on, of stimuli to the raw senses, as we've just seen, play a key role in driving your moment-to-moment brain state – but then so does the personalized *cognitive* element of those inputs (the size of the stone, as well as the force with which it's thrown). And if the size of the stone corresponds to long-term, local neuronal connections, which are in turn driven by the long-term and ongoing environment, then the surroundings in which you spend eight hours of your life each day will be crucial in shaping your consciousness. If unsophisticated rats and mice are transformed at every level by an 'enriched' environment – from the neurons and chemicals in their brain up to complex brain circuitry and, ultimately, their behaviour – then how will the everyday workplace affect the brain of someone like you, a unique human being?

But, this time, if we're thinking about the effects of the complex everyday environment on the human brain, then that environment will inevitably be multifaceted and multisensory: we can't really regard enrichment as the same monolithic, all-or-none lifestyle that's just fine for experiments on rats. The paradigm of environmental enrichment can be easily created and used as an experimental set-up for laboratory animals, but the situation is infinitely trickier for humans, because in any study of enrichment it is impossible to contrive a 'differential' control scenario. It would be unethical to shut away even the most willing of volunteer subjects for long periods in a deliberately under-stimulating environment, and certainly not for a significant enough period of their lives to ensure a clear differential for comparing them with individuals who are to be enriched.

Sadly, the nearest approximations to such a dramatic environmental tampering with the human brain were the tragic victims in the Romanian orphanages during the brutal Ceausescu regime. Due to a rising birth rate resulting from the prohibition of abortion and contraception, many children were abandoned in orphanages in the 1970s and 1980s, along with those who were disabled and mentally ill. These children suffered institutionalized neglect and abuse, and the terrible deprivation clearly had an effect on their brains – both their mental and their physical development.[3] Many experienced significant delays in fine motor control, language and socio-emotional functions.[4] In one survey of Romanian orphans between the ages of twenty-three and fifty months, all exhibited deficits in their mental processes which were, surprisingly, not related to length of time in the orphanage, age at entrance, or birth weight. The sheer horror of the environment itself seemed to trump any more subtle variations.[5]

Doctors studying the orphans observed a decrease in both white and grey matter in the brain, as well as extensive under-use of glucose (a sign of neuronal under-activity) in a wide range of brain areas. They also detected an abnormality in a key aspect of brain function: neuronal connectivity.[6]

The children were most proficient at social interaction with each other, since that was the one kind of activity they had been allowed. Moreover, some of the worst effects of the deprivation were ameliorated when they were placed in foster care,[7] which illustrates yet again the ceaseless adaptability of the brain to the environment, the way in which it reflects every moment of experience and normally thrives on incoming stimulation.

Of course, most human experience could not readily be classified into polarized conditions of enriched or otherwise. Even an extreme, literally impoverished lifestyle characterized by living on the breadline and sleeping on the streets could well contain elements of stimulating interpersonal interaction and memorable and instructive experiences, be they good or bad. More subjective factors will clearly play an important part, such as personalized reactions to, and idiosyncratic interpretations of, particular events, along with each person's individual agenda and motivation: moreover, there will be the most diverse range possible in the particular kinds of environment that each human being might find enriching. As Dr Diana Frasca from the Toronto rehabilitation unit comments: 'Enrichment, therefore, reflects environmental complexity (the opportunity to participate in different sports, clubs or social networks, and to engage in intellectually demanding activities), one's inclination to participate in the environment, and the frequency of participation.'[8]

The other less obvious, yet hugely significant, difference between studies of environmental effects on humans compared to rats is that rodent outcomes will be measured along a single scale, be it upregulated, increased levels of chemicals, or more neurons with more connections or, eventually, perhaps, a behaviour in which the animal is more proficient in a certain test than the controls. Not so with humans. Not only will each individual react in a highly individual way to an environment but, in turn, it will be far from obvious how to evaluate the unique effects of that environment. For instance, willingness to eat novel food is a good performance measure in rats when testing for levels of anxiety: it would be hard to imagine using this as a credible means of evaluation in humans, not least as the opposite pattern might be seen: stress-induced overeating.

In order to take on the daunting task of exploring the complex effects of the environment on human consciousness, the workplace is a very good start: it offers, after all, a relatively more standardized lifestyle and, arguably – with its explicit goals, rules, hierarchies and agendas – a slightly simplified version of the real world. Yet we'll still need to look beyond the mere *physical* features of this environment and consider the way that we interact with it, moving around in open space. Each individual's *subjective reaction* will play an important part in shaping assemblies, and hence consciousness. As such, the workplace offers an intriguing opportunity to explore the more sophisticated aspects specific to the human mindset: identity and self-consciousness. So, where to begin? Perhaps with the most obvious: the objective, physical features around you, of which the most ubiquitous of all, even in an office, is colour.

PHYSICAL FEATURES: COLOUR

A human with normal vision is able to appreciate a staggering 2.3 million discernible colours;[9] and the physiology of the way in which colour is processed in the brain has been extensively researched and documented.[10] Still, the nagging mystery remains as to what exactly an individual experiences subjectively when they see colour. We can only describe our perception of say, red, by making comparisons with various red-coloured objects – apples, blood and the like. The experience of seeing red is therefore frequently cited as an example of 'qualia', the elusive quality of a subjective conscious experience that no one else can share or experience first-hand.[11] Perhaps colour is so often chosen as an example of qualia because it is the most basic element of a conscious experience.

However, in the unlikely but hypothetical instance that all you could see in front of you was simply a sheet of an unbroken, homogeneous colour, then it is still possible that simple experience could play a key role in the generation of assemblies, and hence consciousness, at the most primitive level of all: sheer excitement. Recall that one of the important factors in determining the eventual size of an assembly is degree of arousal, which is in turn determined by the chemical fountains of dopamine, noradrenaline and other chemical siblings. It turns out that merely the wavelengths of different colours can affect your arousal levels differentially and directly, simply because of how close or otherwise objects with different colours appear to be (in turn, all down to their different wavelengths).

Two hundred years ago, Goethe described blue as a 'receding' colour, in contrast to red, which is 'piercing into the organ'. The yellow wavelength is relatively long and fundamentally exciting: various psychological studies have shown that a yellow stimulus provokes an emotional reaction and therefore can lift spirits. In contrast, green activates the medium-wavelength spectra of the colour-sensitive cells (cones) in the central retina.[12] Some scientists have therefore suggested that green light, since it is at the centre of the spectrum, strikes the eye in such a way as to require no subsequent ocular adjustment whatever and is, therefore, arguably more readily restful and relaxing.

Moreover, when one colour is contrasted with another, this also affects arousal levels. Red has a longer wavelength than blue and it is thought that *differences* in the wavelengths of two colours can make two adjacent squares of those colours seem to be different distances away.[13] So, colours in and of themselves, in the most abstract way possible, will affect human consciousness, simply due to their physical properties, rather than by virtue of any psychological 'cognitive' associations they may accrue later in life.

What might be happening at the level of the neurons? By directly affecting arousal differentially, different colours will be changing the levels of those various modulatory fountains of chemicals in the hub of the brain. These all-pervasive chemicals will, as a result, consequently modify how readily assemblies can form by determining the viscosity of the puddle: the ease, or otherwise, with which the ripples can be generated through the puddle water is a key consideration. Thick, stagnant liquid will be a less efficient conducting medium, slowing down the spread of the ripples, compared to water that might be completely clean and clear: so, the family of modulating chemicals emanating from the hub of the brain could determine the ease with which an assembly is recruited, and to what extent, all depending on how, literally, exciting or otherwise a particular colour might be.

Then there is the force of the stone throw. A colour of a fixed wavelength will still vary in brightness: a pale red patch is not going to be as stimulating as a vibrant, dazzling one, even though both are red. So, already, a colour can trigger consciousness in two ways, by virtue of its mere psychophysical properties: firstly, the wavelength, which determines the level of arousal through the various modulators (the viscosity of the puddle water) and, secondly, the degree of brightness (the force of the stone throw). What now could determine the size of the stone, namely the particular significance of the colour?

Unsurprisingly, there are characteristic human responses to colour that wouldn't feature at all in the repertoire of environmental-enrichment studies in animals. For example, the spectral quality of light can modulate individual emotional brain responses in humans, as revealed by brain imaging. Investigators exposed subjects to emotional vocal stimuli while they were undergoing fMRI scans in alternating forty-second periods of green and blue light.[14] The individuals showed greater functional connectivity in brain networks linked to emotional experiences (the amygdala and hypothalamus) when exposed to green light and greater responses in another area (the hippocampus) under conditions of blue light. Because the hippocampus is important for the laying down of memories, the researchers suggested that blue light promotes heightened emotions and memory processing to encourage rapid brain reactivity to emotional challenges and, in turn, aid the organism's adaptation to its environment. Such an interpretation is probably very simplistic – after all, we've seen that brain regions are not autonomous mini-brains – but, at the very least, this study shows that different colours can affect differentially how the brain responds to environmental stimulation by virtue of their respective cognitive effects.

In real life outside of the lab, it is virtually impossible to tease out such direct effects of pure colour – the force of the stone and the viscosity of the water in the puddle – from the deep-seated associations each particular colour will trigger: the *size* therefore of the neuronal stone. Many of the effects of colour will be mediated by learned associations built up between the colours and personal environment: the post-natal neuronal connectivity that we've already discussed will be such an important factor in determining eventual degrees of consciousness. For example, if it is well-established that red boosts attention, it is because it is associated with, say, the traffic-light signal to slam on the brakes, as well as with other signs of prohibited entry and all manner of danger and hazards. However, if the colour was chosen in the first place for this signage precisely because of its direct physiological effects on pulse rate, heart rate, and so on, it will be impossible to dissect the sensory from the cognitive: the effects of the increased arousal level via the modulators will have a reciprocal impact on the associations of a certain colour, so that the two are mutually reinforcing.

In the case of red, and operating under the assumption that the colour does indeed, for either innate or learned reasons, boost attention, the overall effect would be to enhance performance in cognitive tasks that involved attention to detail ('focused attention'): yet, at the same time, an individual might as a consequence be more vigilant, on edge and risk averse and, therefore, less motivated. Performance would decline. After all, we pay attention not just to interesting or potentially beneficial stimuli but to ones that might cause us harm; when these appear, we are, above all, reactive to risk, rather than motivated to be proactive. Meanwhile, imagine now that the immediate environment around you contains plenty of the colour green, playing to our most primitive lifestyle of the forest, as well as suggesting the presence of water. At the most atavistic level, we are reassured by the sight of green. Then again, on a more negative note, green can convey blandness, since it is the most ubiquitous colour in nature. By the same token, blue is readily associated with the ocean and the sky and therefore evokes feelings of openness, peace and tranquillity.

Following on from such cognitive connections, it is well-known that colour influences consumer behaviour.[15] The most commonly used colour for the logo of major companies – Facebook and Twitter, for example – is blue; this colour is in turn linked to competence.[16] Blue, due to its associations with water, is most appropriate for goods that are functional, while red, which is associated with status, is the colour of choice for luxury goods, such as sports cars.[17] And it is this single colour alone that has been by far most intensively investigated, ever since the British physician Havelock Ellis wrote in 'The Psychology of Red' (1900), 'Among all colours, the most poignantly emotional tone undoubtedly belongs to red.' From the many life-and-death connotations of blood to the flushed face betraying basic emotions of anger or embarrassment to the colour of ripe fruit, red has more 'significance' (and therefore potentially more neuronal connections) than most other colours.[18]

But perhaps the still more fascinating feature of colour as part of the environment is not so much how it inputs into your consciousness but the ensuing *type* of responses which might then result. For example, red is preferable if the task on hand requires an individual's vigilant attention – such as memorizing important information or understanding the side effects of a new drug – as it activates an avoidance reaction that might be particularly appropriate.[19] But red would not be the colour of choice if you are striving, either at work or at home, for an environment that promoted a different type of behaviour altogether: creativity.

In one experiment at the Sauder School of Business in Vancouver,[20] the potential effects of colour on human creativity involved setting a battery of different tests that assessed a range of psychological attributes. One was a simple test of word recall with screen displays using different colour backgrounds; words shown on a specifically red background prompted participants to recall more words afterwards. A proofreading test asked participants to compare sections of text and indicate whether they were identical or not; once again, the instances with a specifically red backdrop on a computer resulted in a better performance, as a measure of accuracy. Then, more significantly, subjects underwent a test in which they were asked to list as many uses as possible for everyday objects, with their responses given a score based on number and ingenuity: this was an evaluation of creativity. It turned out that background colour during the test didn't influence the number of responses but it *did* improve their ingenuity, namely, the 'creativity' of the answers offered; this time, it was a blue backdrop that produced the more imaginative responses.

The scientists conducting the study also gave participants the

Remote Associates Test (first introduced in Chapter 3), in which three words are given, all of which are related to a fourth word, which the participant must provide, for example, 'shelf', 'read' and 'end' are all related to 'book'. Interestingly enough, it was again words presented against a blue background that produced more creative answers than those against a background of red. As these studies suggest, if a task calls for creativity and imagination, such as designing something, coming up with an idea for a new product, or having a brainstorming session, you should seek out a blue room; on the other hand, red environments are best for recall and attention to detail.[21]

Blue is close to green in the colour spectrum, and may also promote creativity by giving the individual mind the opportunity to 'roam free', at least metaphorically, in a natural environment of green trees and blue sky. In a telling investigation, scientists used a number of different ways to compare the effect on creativity of the colour green with that of red, grey and white.[22] The study gave participants a variety of tasks: first, to list creative ways to use an everyday object, then a test known as the Berlin Intelligence Structure, in which participants are asked to draw as many objects as possible based on a geometric figure, and, finally, the Instances Task, in which participants are asked to provide as many examples within each of four different categories (for example, 'things that are round') within a set time. The investigators found that the colour green was of no benefit in non-creative, analytical tasks but it did help participants perform better on creative tests than red, white or grey. The researchers concluded that blue was not *the* creative colour and suggested that lightness, chroma or some other interaction between the three primary colours were the real factors that affected the final outcome.[23]

These conclusions may all seem far-fetched or simplistic. But if certain colours *are* linked to certain types of responses, this is probably because they influence the three distinct interactive factors that determine the final degree of consciousness – the extent of the ripples in the puddle: firstly, the size of stone (the degree of pre-existing associations); secondly, the force with which the stone is thrown (the degree of luminosity and brightness); and, thirdly, the ease with which ripples can be generated, that is, the viscosity of the water (the availability and amount of modulators, which is in turn linked to the general degree of arousal).

INTERACTION: OPEN SPACE

Along with colour, there is one feature of the environment that is so obvious it seems almost fatuous to mention it: empty space. While perception of colour may well be a passive, one-way experience, we have just seen that particular colours can make us more vigilant or creative. The space around you invites not just a reaction but an *interaction*: the possibility of physically moving around. The experience of moving around in defined spaces will in turn feed back into your brain, modifying your neuronal connectivity and hence changing the size of the stones, and hence of the assembly and, ultimately, of your prevailing consciousness. So what exactly is this special space called an office?

Since Roman times, the loose concept of 'office' has featured in daily life ('official' administrative business is carried out, either by an individual or a group): however, initially, the emphasis was on the position held by a person rather than the place – as in the 'office' of a minister. Over subsequent centuries, society ceased to be controlled and run by complex bureaucratic government. It was only in the eighteenth century, with organizations such as the East India Company, that the notion of an office was revived – but this time as a much-needed physical space. The office really came into its own at the start of the twentieth century, because of a combination of different, yet timely, technological inventions: new electric lighting allowed for greater productivity than expensive gas lighting or many windows ever could; typewriters and calculating machines sped up information handling; telephones and telegraphs enabled office buildings to be independent entities with reliable but remote lines of communication; and, finally, with the invention of the lift and steel-frame construction the familiar high-rise buildings that started to dominate the urban skyline were made possible. Inevitably, over time, office architecture evolved and became the focus of design experimentation.[24] The high-rise boxes that became popular after the

Second World War were later thought to promote a sense of anonymity and inflexibility in workers.

The rationale since the start of this century has been to encourage a new way of working by rethinking physical space so that it meets human social needs. In *The 21st-century Office* (2003), Jeremy Myerson and Philip Ross suggest that design should be determined according to four key terms. The first is *narrative*: in stark contrast to the visual uniformity of offices of old, the contemporary space is supposed to tell some kind of story about what a company stands for. Take the headquarters of the smoothie company Innocent in London, where the lower floor is covered in AstroTurf and the mezzanine is dotted with American diner-style booths; it even features an old red phone box. The second term is *nodal*: new office spaces often contain different sections, each of which fulfil the various needs of the company and of the employee, with flexible meeting spaces, communal relaxation areas, and so on. The third is *neighbourly*: an environment that encourages interaction between employees; and the fourth is *nomadic*: the workplace no longer needs to be in a fixed space and neither does work have to be done at a particular time, as digital technologies enable greater flexibility in working styles.[25] The general idea is to promote among employees a greater sense of freedom in terms of time and space. But has this new approach been successful in achieving what it set out to do?

Research into the impact of open-plan office design on performance is divided and contradictory. On the one hand, open-office spaces can erode well-being, due to loss of privacy. On the other, the exact opposite has been argued: that they improve job satisfaction, staff morale, the exchange of information and productivity. Another consideration is of course financial: open-plan design is often about financial pressures – fitting more employees into smaller spaces, in response to rising rents. Interestingly enough, even at the offices of Google, a company often held up as a 'beacon that should be leading other organizations away from the traditional',[26] the open-plan, recreational-style offices are reserved for departments only where *innovation* is the primary goal. So, might open spaces only be most appropriate specifically when the intention is that they lead to open minds, which are in turn more likely to be creative?

Two case studies featuring companies which tried to promote creativity in their office space give us some clues.[27] The first took place at the headquarters of a pensions specialist in Sweden, where the organization moved away from small, traditional offices to more space-efficient areas with 'hot-desks', work stations that belong to no one on a long-term basis but can be used by anyone. The BBC has also recently switched over to 'hot-desking', which, reportedly, has led to a lot of discontent among employees.[28]

In the Swedish case study, all personnel were located in open-plan offices, including the CEO and the managing director. Comfortable kitchens with sofas and modern appliances were installed on every floor. As one manager optimistically concluded: 'When employees sit together in a shared space they are in the middle of the information flow. They are constantly exposed to what is going on, and they can speak more freely with each other.'

However, it emerged from interviews and objective observations that spontaneous interaction, creativity and learning were not an automatic outcome of the new open-plan design. While the staff admitted that it was indeed easier to see the people you needed to talk to, loud discussions nearby were distracting. Meeting rooms became booked up quickly. Another reason why open-plan offices were unpopular was that employees felt as if they were being monitored all the time, by other workers as well as management: some felt that, due to a lack of privacy, they were obliged to maintain a false work identity. In the end, many staff took advantage of flexitime opportunities and would work at home more often. So much for open-plan offices encouraging 'spontaneous interaction and creativity'.

The second case study was at a UK call centre that was centralized into an open space facility dubbed 'The Dome', and aspired to combine sensations of 'recreation, creativity and vibrancy'. Activity rooms were included in the plan, such as a sports-themed room with pinball, table football and so on, as well as a 'Mediterranean' café. A lack of fixed boundaries was key to the design of the space. Hot-desking was encouraged, though, in practice, this didn't happen much, as competitive teams formed naturally and remained together, even personalizing their spaces and moving furniture around to mark their territory out more clearly. Some employees found the environment 'wearing' and objected to having such little personal space. Others claimed that the organization was more fragmented and that inter-group boundaries and differences were even more conspicuous than in other organizations they had worked for.

These two studies suggest that open-plan office designs often lead to feelings of greater surveillance and perceived control. This kind of design ultimately resulted in a kind of auto-surveillance whereby employees disciplined themselves, and in ways that frustrated rather than facilitated spontaneous interaction, fun and learning. Rather than making employees feel more creative and flexible, it made them, in other words, more self-conscious.

Open space isn't simply an absence of clutter but, in itself, its emptiness can be potentially frightening: a well-established model for evaluating anxiety in rats is to see how readily they move around in an open-space arena.[29] Of course, rats aren't going to be concerned with subtle aspects of work performance – but a much more basic element may be common to both humans and rodents: exposed spaces make both species feel vulnerable. For humans in an office, this vulnerability might translate to a fear of surveillance and a feeling of loss of privacy – both of which are connected in turn with identity. After all, inherent in the notion of privacy is *privation*, the shutting out of other people by, say, drawing the curtains. Once this firewall is erected, you, the private individual, are protected in a way that wouldn't occur were you to be endlessly downloading every thought and perception with others: if those thoughts and perceptions are an integral part of who you are – your identity – then lack of privacy in an open-plan office might erode your sense of identity. Such an office could therefore offer not so much freedom to be yourself but the disconcerting opposite. When we think about how the immediate space and environment around us affects our consciousness, we need to probe well beyond their physical features to the more psychological ones.

Way back in 1927, the Western Electric Company, based in Chicago, conducted a seminal investigation into the impact of improved lighting on the motivation and productivity of twenty thousand employees at one of their factories.[30] Perhaps not surprisingly, productivity increased in the group that worked in a brighter workplace, but a baffling and intriguing finding was that performance also improved in the control group that continued with the same, lower lighting levels they had always had. Crucially, the key factor was that the control group were informed that they were part of a psychological study, and therefore worked harder, knowing that they were the subject of scrutiny. The study revealed that workers were motivated not so much by their physical conditions, or even economic factors, but instead by 'emotional' issues such as feeling involved and being listened to.

So what potential might there be for the kind of an environment, somehow based on these intangible 'psychological' factors, that could maximize individual well-being? One investigation from the College of Architecture and Environmental Design, in California, focused on a group of people who were expected to think extremely creatively in their work: scientists in various different university research centres.[31] Perhaps inevitably, when the scientists were given more freedom of movement and access to face-to-face consultations, it made it easier for them to communicate and, in turn, this interaction sparked more imaginative thinking and innovation. In general, layouts that brought the scientists closer together, made them more visible or 'more likely to bump into each other', increased overall innovativeness. Similarly, in another study, it turned out that a staggering 80 per cent of 'consultations', defined as face-to-face conversations, occurred through meetings that were not pre-planned but were instead unscheduled office visits or accidental encounters, and these kinds of interactions were universally preferred.[32]

Having a conversation with a colleague can be enriching on many levels: the simple pleasure of human eye contact; the warm glow from feedback that what you are saying is worthwhile as you register the other person's encouraging body language; above all, the new ideas that the colleague may plant in your mind that lead to further original thought and fresh implications. Then there is the reinforcement of your unique identity when it is asserted, realized and reaffirmed in a real-time conversation. None of these effects can be readily measured, yet they highlight further the 'psychological' features of a working environment that can be set in train only indirectly, by designing spaces that would ideally allow for merely the possibility of more conversations.

Yet we've seen that open-plan offices can jeopardize creativity or the well-being of employees. The ideal design therefore seems to be an environment that allows for individual privacy and security but also offers the opportunity, if the person chooses, to venture out into shared spaces and have chance encounters.[33] Unlike the environmental enrichment paradigm in animals, the impact of the workplace on employees is essentially a two-way street: while rats may be the straightforward and psychologically unencumbered recipients of whatever comes their way, people have considerable personal baggage. Subjective and emotional factors play a far greater role, and hence individual variations in both initial expectations and subsequent responses are only to be expected. But since the environment – specifically, the working environment – has such clear effects on so many different levels, be it arousal, perception, memory, well-being – the next step is to see how these different factors in the environment might converge on individual consciousness: more specifically, your self-consciousness and the subjective feeling of who you are.

SUBJECTIVE REACTIONS

Unlike studying the impact of the environment on animals, where the outside world, almost literally, leaves its mark unconditionally on an obliging, receptive brain, a study of the effects of the environment on human thought and behaviour brings into play a much more elusive concept which is, in turn, either expressed or threatened by the space around you: identity.

I would argue that there are five criteria necessary for there to be a particular, personal identity:[34] first, being conscious; second, having a mind and, with it, a value belief system; third, acting on those values in a specific context; fourth, eliciting a reaction with those actions; fifth, embedding that particular instance of action–reaction into a wider and more generalized life-story. Let's look at each specific step in more detail.

First, rather obviously, you can be aware of having an identity only when you are capable of awareness. So, you need to be fully conscious, that is, neither asleep nor anaesthetized. Second, your 'mind' has to be fully operational – in other words, not 'lost' under the influence of drugs or alcohol, nor distracted by an ongoing fast, sensational experience, as we discussed earlier, of 'letting yourself go'. You need to be able to experience a personal interpretation of what is happening at any one moment. Your unique mind – your personalized connections – will enable you not just to make sense of what is happening but to apply values and beliefs generalized over time from your particular memories that will be impacting on each individual moment of experience. In other words, you must not only be conscious, but self-conscious. Before we examine the further items on the shopping list needed for there to be a full-blown identity, let's pause to look a bit more closely at what is required just to be conscious of what you are doing.

Much neuroscientific study has been devoted, as we saw back in Chapter 1, to the 'higher' states that constitute self-consciousness[35] and to what extent they occur in animals as well as humans.[36] While it may not always be necessary continuously to have a complete sense of self, it seems that an awareness of one's actions can be very important in determining what we do subsequently. Consciousness of what we are doing – conscious knowledge – can be differentiated experimentally from the subconscious by, of all things, a wagering task. Participants in a study were shown individual letters they had to identify for only five to ten milliseconds. They responded with errors at a rate greater than chance: this must mean that the stimuli that are being presented are processed subconsciously. However, when the presentation period was longer than fifteen milliseconds, presumably allowing for 'consciousness', accuracy improved. When offered the opportunity to wager low or high on their choices after these short-duration stimuli, the participants refused to gamble for higher stakes even when they were correct. Only when they were conscious of what had happened would the subjects choose to take a risk.[37]

Yet this operational demonstration doesn't really help us in defining self-consciousness. We've seen already that the subjective experience of 'what it feels like' is dependent on personal associations and memories and thus unique emotional states: yet you are not necessarily conscious that it is *you yourself* feeling them. In other words, you don't always need to be self-conscious – conscious of yourself. So we still need to work out how such self-consciousness would differ from, say, the subjective consciousness of a baby who, nonetheless, doesn't realize they're a baby.

One idea is that self-consciousness differs from these other ordinary brain states because it 'informs the agent of its own internal states'. However, surely this theory commits the Readout Fallacy[38] of assuming there is some kind of spectator inside the brain. Who is this 'agent', if it's not the internal brain 'states' themselves? This suggestion is therefore not particularly helpful in understanding what might be happening within the brain. Another, less vague, theory is the Radical Plasticity Thesis. The idea here is that it is the actual process of learning itself that makes us conscious – the brain not only learns about the external world but somehow has its own representations of itself, known as metarepresentations. This theory links to two computational theories of consciousness. The first is that consciousness derives from 'fame in the brain', which was mentioned briefly in Chapter 1:[39] whatever the state, at any point in time, that has come to dominate information processing, such will be the essence of our consciousness. The second is that the crucial issue of whether or not you are conscious depends on the involvement of metarepresentations.[40]

However, we are still no wiser as to why 'higher order' processing and being self-conscious would be so different, in brain terms, from ordinary old consciousness. As the Radical Plasticity Thesis leads its author to conclude, 'It is because of the fact that one is conscious that one is conscious, that one is conscious!' This is not a particularly informative insight and, moreover, it overlooks the instances in all non-humans, and sometimes even in our own species, notably in babies, where consciousness is presumably being experienced but without self-consciousness. Even when these definitions and descriptions might be semantically helpful in differentiating self-consciousness from consciousness, the distinction is hard to envisage mechanistically in terms of the physical brain.

One possibility is that the difference is not so much qualitative as quantitative: it could be that, during brain development, neuronal assemblies need to pass a critical-size Rubicon in order for self-consciousness to be possible. Still, some irritating questions remain: why would a larger assembly, differentiated only by the *quantitative* feature of size, now acquire a further functional property – a metarepresentation – that is *qualitative*? We can only really tackle this question once we have journeyed throughout your day and reached conclusions at the end of this book. Even so, as it stands, this scenario of the necessity for potentially generating large assemblies could explain why pretty much all animals (with the possible exception of sophisticated primates), as well as babies and infants, are never self-conscious. Moreover, a continuously variable size of assembly would also account for why, from time to time, adult humans with an unusually small assembly are able to *lose* their self-consciousness – to let themselves go. Other theories of self-consciousness do not account for this flexibility.

Identity is an even more complex prospect to understand in terms of the brain. After all, if you were washed up on a desert island, you would have an intact mind and be self-conscious, but your identity of being a specific person recognized as such by others would be harder to articulate. So there must be additional requirements to distinguish self-consciousness from these all-important feelings of identity – and this brings us to the third criterion. When you express your identity, you will react in a particular way, determined not just by your own beliefs and experiences, your self-consciousness, but by the context prevailing at that particular moment, the prevailing role you are playing as a member of a family, a sports team – or, indeed, a workforce. This 'reaction' can of course be imagined rather than always played out in the three-dimensional world.

As a result, your actions will elicit reactions from others, which in turn will modify your memories and how you will react to a given situation the next time around: this would be the fourth step. Ultimately, then, we come to the fifth step: these moments of action–reaction in a conscious present will now be incorporated into a wider narrative of your cohesive past-present-future: your unique life story. It is this subjective awareness of your unique life story *at any one particular moment and in one specific context* that constitutes the immediate 'feel' of your identity.

We've seen repeatedly that the state of losing your mind is typified by a lack of self-consciousness. So, to attain 'mere' self-consciousness would require going only as far as having a functioning mind. However, in order to enjoy a sense of a highly specific identity, something additional must now occur: you need to place yourself in a context (third criterion), elicit reactions (fourth criterion) and embed those actions and reactions in a still-wider context (fifth criterion). In terms of the brain, we need to understand how our extensive neuronal connectivity accounts not only for the first two steps of being conscious and being self-conscious but also the third, fourth and fifth steps listed above – the way we create our unique identity.

As well as having a functioning, self-conscious 'mind', you'll now need to be involved in a specific situation in which you're playing in a team, singing in a choir, eating a meal with your family or having a discussion at work – a situation in which you have an immediate role to play. In it, your words and actions will evoke responses from those around you, which will in turn determine how you'll respond subsequently, perhaps even modify your wider views and beliefs. And, in however infinitesimal a way, this particular one-off experience will build itself, like the proverbial brick in the wall, into the complex life story that gives you your overarching identity. These iterations that characterize how you incorporate a specific context and role into the complex individual you feel you are, will be based on ever more complex connectivity of neuronal networks. As such, we could imagine an ever bigger stone and hence far more extensive ripples in the puddle – an ever deeper consciousness as you feel a strong sense of being unique in a unique context.

One of the consequences of feeling a strong sense of identity is the potential for the expression of that identity through some form of creativity. In Einstein's words, 'It is important to foster individuality, for only the individual can produce new ideas.' So let's return to the context of the office and think about how these environments can foster a strong sense of individual identity, something which helps people be truly creative. Creativity and creative thinking are at a premium in a vast range of diverse organizations: yet they cannot be simply downloaded nor bought, nor can they in any way be guaranteed. It's no wonder that employers are very keen therefore to ensure that at least the opportunity for this most prized commodity is maximized in the workplace. Since a stable identity is so important to well-being,[41] and since we've seen that a crucial factor in our daily working environment is the way in which different types of these environments can either threaten or help us express it, it is not surprising that much thought is being given to designing work environments where employees can 'be themselves'.

So how might a working environment promote creativity as the ultimate realization of the individual? It's only to be expected that some people will, from the get-go, be far more creative than others. Although, therefore, it might seem unrealistic to be prescriptive, the psychologist Øyvind Martinsen recently drew up a list of creativity characteristics by comparing students of marketing and artists with managers from less creative industries[42] – and seven essential qualities of the creative mind emerged.

The first was associative orientation: according to the study, the creative individual is imaginative, playful and has a wealth of ideas. Secondly, need for originality: the individual must resist rules and conventions; they have a rebellious attitude, due to a need to do things no one else does. Thirdly, motivation: they have a need to perform and have an innovative attitude and the stamina to tackle difficult issues. Fourthly, ambition: a hallmark of the creative personality is a need to be influential, to attract attention and recognition. Fifthly, flexibility: the creative has the ability to see different aspects of issues and come up with optimal solutions. Sixthly, low emotional stability: perhaps surprisingly, the individual will have a tendency to experience negative emotions, greater fluctuations in mood and emotional state, and failing self-confidence. Seventh, low sociability: creative types can lack consideration for others, often finding flaws in people and their ideas. A final factor in determining creativity could be cultural diversity. Research shows that subjects who had spent time living and learning abroad and who had therefore been subjected to intensive foreign cultural experiences, exhibited more creativity than those who had not.[43] In part, this was because they could understand more readily that the same problem can have multiple solutions.[44]

But, rather than just leaving an individual to be as creative, or otherwise, as their natural tendencies and talents dictate and hoping for the best, some organizations take a more proactive approach by trying to shape the working environment to encourage these natural propensities in their staff. It's not so much offering a wide-open space itself, which could actually be counter-productive: the important ingredient might not be the actual environment so much as how it could be used, to 'play'. Recently, open-plan office design has witnessed a shift towards the promotion of 'fun', 'spontaneity' and 'creativity' through design.[45] Play has been defined as 'the exploration of the possible'[46] and recruiters at NASA and the Jet Propulsion Lab (JPL) found that the best and most innovative candidates were those who 'tinkered' in their youth – in other words, individuals who were curious enough to take things apart to discover how they worked.

The blurring of traditional boundaries between work and play in the office started back in the 1980s. The psychologist Peter Fleming, currently Professor of Business and Society at the Cass Business School, has flagged just how unusual it is for such a fad to last as long as this one has. However, when he conducted a field study of a call centre in Australia in which a culture of fun was very much incorporated into the company ethos, around half of the workforce expressed some degree of cynicism about the company's initiative, and no consideration was given as to whether a culture of fun was related to any kind of increase in productivity or innovation.[47] So not only is there little evidence that unconditional and unconstrained 'play' leads to any type of creative outcome, but such an apparently attractive opportunity might even be counter-productive.

Although this area could well benefit from more research, another approach has been more specific: to explore the power of everyday metaphors such as 'thinking outside the box' on creative problem-solving and cognition – literally.[48] Psychologists have identified two different kinds of thinking: convergent and divergent. Convergent thought refers to intellectual endeavours which involve finding the single best answer. In one study, the participants' convergent thinking was assessed in one of three conditions: literally sitting in a five-foot-square box, sitting outside the box, or sitting in a room without a box. Participants were also given the Remote Associates Test, which revealed that convergent thinking was not changed by sitting in a box, or by being in a room without a box but that it *was* improved when participants sat outside the box. Perhaps they felt less inhibited when the fact that they were not trapped in a box was visible.

In contrast, divergent thinking produces many different solutions to a problem and features both fluency (the total number of ideas) and flexibility (different kinds of ideas). In a study to test this more creative form of thinking, participants were given Doodle and Lego tasks. In the Doodle task, participants had to provide captions for two ambiguous or riddle picture doodles after they had been given an example caption for each: the measure taken to form a result was the amount of deviation of the participants' answers from this earlier example. In the Lego task, participants were shown several Lego models and had to write down up to eight objects which the models represented. The experiments tested the originality of the responses based on the infrequency of certain ideas when participants' answers are studied as a whole.

Rather than present a literal box, this time the effects of the environment on divergent thinking was tested by comparing the effect of sitting, walking so as to make the shape of a rectangle, and free walking. Results of both the Doodle and the Lego tasks revealed that those who were in the group that was freely walking provided more original answers than those seated or walking in rectangles; the answers provided in each of these situations didn't differ.

So, when a subject is physically seated literally outside a box, problem-solving (convergent thinking) was improved: moreover, a period of walking improved originality (divergent thinking) when compared to sitting down, and when walking freely rather than pacing out the rectangular shape of a large box. The researchers deduced that the outside-the-box condition may 'help eliminate unconscious mental barriers that restrict creative cognition' and that 'embodying creative metaphors appears to help ignite the engine of creativity'.

One final factor which encourages creativity might seem surprising: boredom. Could being bored actually make someone more creative? In a recent conference presentation at the British Psychological Society, one of the speakers, Sandi Mann, raised an interesting point which challenged the much-vaunted importance of interpersonal stimulation and enrichment to creativity.[49] He set up two experiments: the first involved participants reading and copying out numbers from a telephone directory before performing a task assessing their creativity, in this case, devising different uses for polystyrene cups. A control group performed the creativity test without preamble. Interestingly enough, the participants who copied out phone numbers performed better in the creativity test. One possibility is that this might be because the directory task allowed the participants' minds to wander and daydream, and that this in turn was beneficial.

To explore this idea more directly, the second experiment required the hapless participants simply to read numbers from a telephone directory: the rationale was to have a more passive and monotonous task that might encourage even greater daydreaming. This group was then compared to those actually copying out the numbers, as well as the controls who were not set either of the boring tasks. This time, the participants who read the numbers showed greater creativity than those copying them out who, in turn, showed greater creativity than the controls. This fascinating observation really does suggest that the kinds of activities that boost creative thought processes are ones that do not demand fast reactive responses and neither need they be exciting or stimulating – rather, they allow the mind to wander.

A THEORY OF CREATIVITY

Interestingly, another recent study has revealed that distraction does aid the creative process.[50] Perhaps allowing for the mind to 'wander' in an undemanding, natural environment does promote creativity. After all, it is a cliché to say you have your best thoughts in the shower, yet this daily routine is often the one time and place where you are guaranteed an opportunity to let your mind range unfocused. Another obvious point about the outdoors is that there usually aren't many people there – perhaps even no one else at all – and, as we've seen, privacy is the other side of the coin to identity, which, as Einstein asserted, is essential to creativity. So, let's assume you have a robust sense of your own individual identity and are placed in an environment free of boundaries and constraints in both time and space, how, then, is creativity achieved?

To recap: by accessing your personalized neuronal connections, and thereby using your mind, you are able to appreciate the world beyond the face value of the senses. Remember the wedding ring, which may mean an enormous amount to, say, a widow, but be merely a gold, shiny object for a young child? As we've seen, your neuronal connectivity allows you to appreciate meaning and symbolism, to see one thing as standing for something else that could never be inferred from the sensory features of the object alone. Sometimes, we make inappropriate or excessive associations which give an over-interpretation of an experience or object, seeing a meaning in it that most others would deem neither realistic nor accurate, perhaps even a little crazy. Discerning faces in cloud formations, or attributing the property of 'being lucky' to an object are everyday examples of such idiosyncratic, somewhat flimsy, associations. Similarly, the pairing of two otherwise unrelated events might seem to some to be a silly superstition but to others a deeply significant 'sign' or portent.

Not only will your neuronal connections allow you to imbue objects, people and their actions with your own personalized significance, they will enable you to understand and navigate the world as you live in it. Imagine someone appearing before you dressed in a ghost costume: thanks to the checks and balances of your established neuronal connectivity, you can make sense of what you are seeing and not have to take the bizarre apparition at face value. A small child, or someone suffering from dementia, who lacks these connections, will be at the mercy of the here-and-now sensory experience alone: as a consequence, they might well be frightened, disorientated and confused at seeing something they can't 'understand'. Since this personalized neuronal connectivity is the physical basis of our individuality, and if individuality is a prerequisite for creativity, this might help us develop a theory as to how this ultimate expression and fulfilment of an individual might be achieved. We can differentiate three stages in the creative process, each of which will be expressed in terms of neuronal connections.

Let's call the first stage *deconstruction*. An essential feature of creativity is to challenge dogma, whether it be the location of the nose on the face (witness Picasso), where the engine fits in a car (witness the design of the Mini) or why ulcers need not be related to stress (the ground-breaking theory of Barry Marshall).[51] In physical terms, such a process would involve the inhibition or overriding of a pre-existing connectivity between brain cells, dismantling the previous understanding, which could sometimes be regarded as bias, prejudice or dogma.

The second stage will be *new associations*. All creative work, whether it is artistic, scientific or a simple but novel domestic insight, involves bringing together new elements, be they words in an original combination, colours and shapes in an innovative form, or the linkage of previously independent facts leading to a fresh idea. However, although this stage is necessary, it is still not sufficient: just think of a child's drawing, or a schizophrenic's poem,[52] which might bring together unusual combinations of shapes, or words, without being regarded as a true example of creativity or original insight.

The essential third and final stage will be one of significance. This key requirement is for an all-important link – the new association – be it in physical shapes, musical sounds or scientific theory, to have *significance* to oneself and to others: not only should a novel association be forged between neurons but it should trigger further associations and connections that give the work meaning. The more wide-ranging the associations driven by this new connectivity, the deeper the meaning will be: the greater the 'Aha!' moment.

So, what *is* the optimal environment to foster these steps in creative thinking? All the three stages above are based on one essential prerequisite: confidence. Along with confidence comes a sense of well-being in which you're not necessarily reacting to others but enjoying being unconditionally yourself. The shower is one wonderful example of this kind of environment.

But aside from the shower, what would the ideal inspirational space look like? Of course, one size will not, literally, fit all when it comes to catering for individual thought. But the impact of the environment on the brain and, in particular, on the human brain, is irrefutable. Various broad categories will be critical in designing a workplace environment to promote the best possible individual thought processes. Obviously, features of *the physical space itself* will be key, such as certain colours or photographs, or windows looking out, ideally, on to natural scenes (as we saw in Chapter 3 how landscape can aid thought and creativity). Then there are features catering for

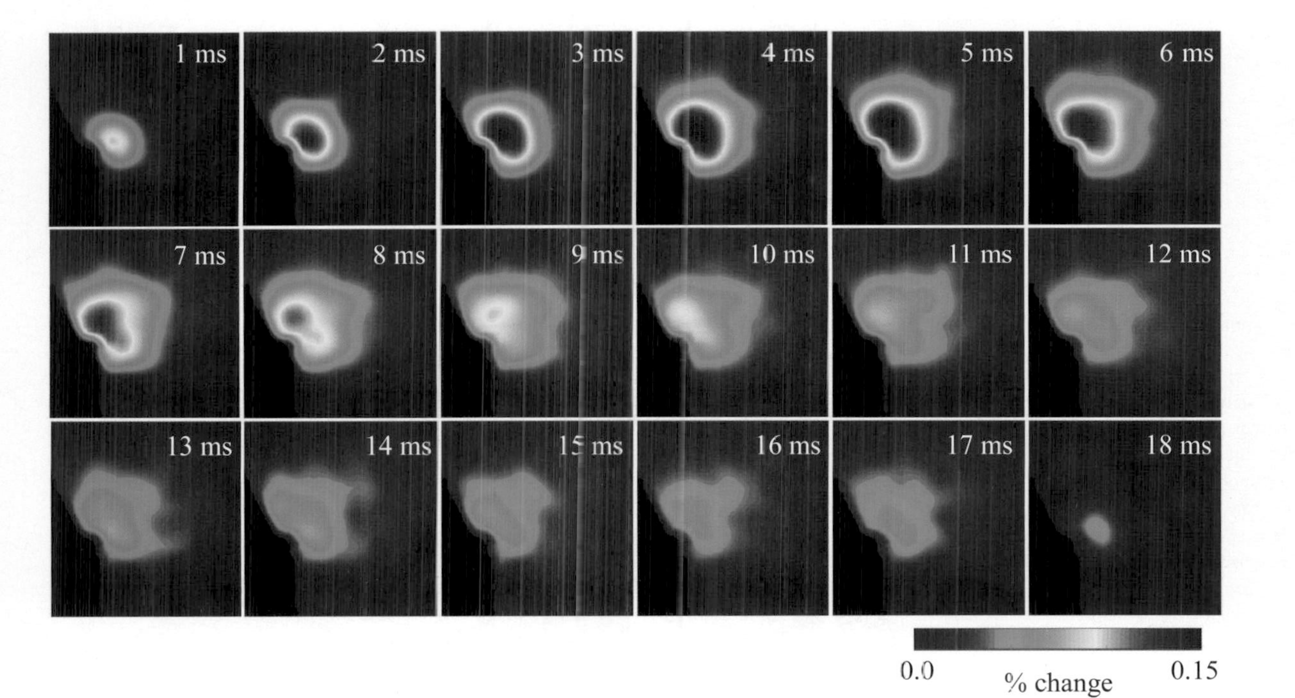

Plate 1: Visualization of an 'assembly'. Sequence of images one thousandth of a second apart, showing widespread activation in a rat-brain slice following an initial pulse of stimulation lasting a tenth of a microsecond. The highest activity is at the centre, gradually decaying to the outer limits – a little like ripples in a puddle from a stone throw. (Badin and Greenfield, unpublished).

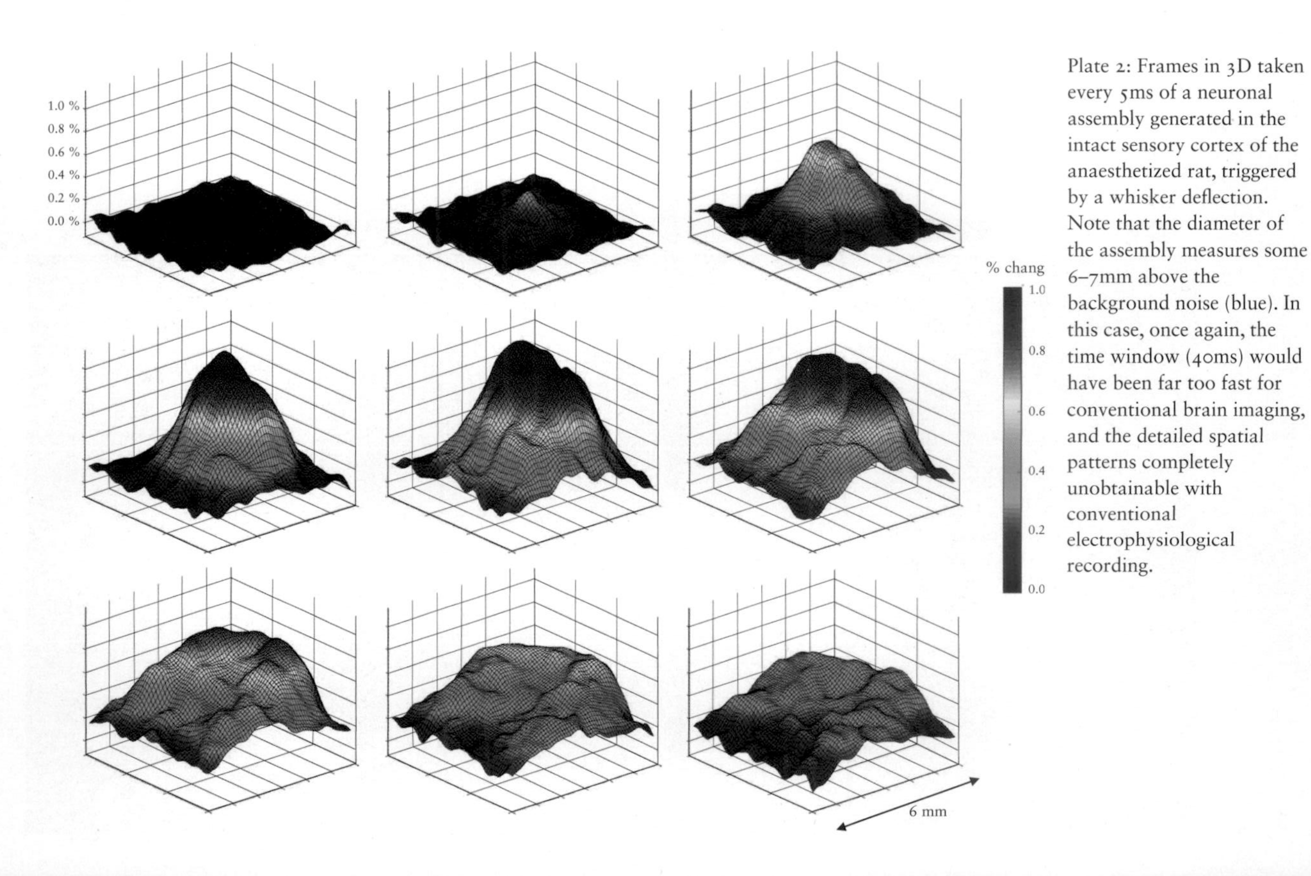

Plate 2: Frames in 3D taken every 5ms of a neuronal assembly generated in the intact sensory cortex of the anaesthetized rat, triggered by a whisker deflection. Note that the diameter of the assembly measures some 6–7mm above the background noise (blue). In this case, once again, the time window (40ms) would have been far too fast for conventional brain imaging, and the detailed spatial patterns completely unobtainable with conventional electrophysiological recording.

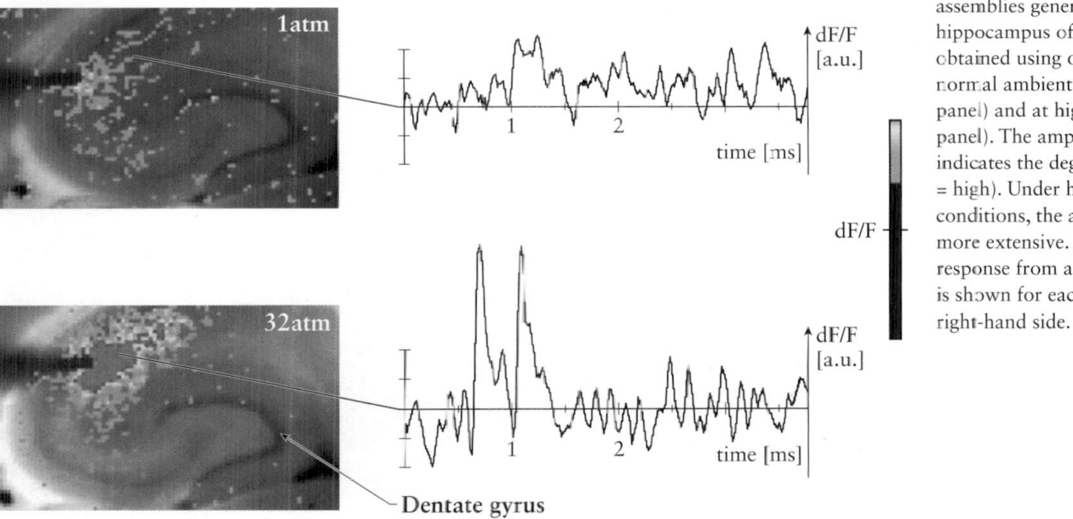

Plate 3: Fluorescent images of assemblies generated in the hippocampus of a rat-brain slice obtained using optical imaging at normal ambient pressure (upper panel) and at high pressure (lower panel). The amplitude of the signal indicates the degree of activity (red = high). Under high-pressure conditions, the assembly is much more extensive. A time course of the response from a small area of tissue is shown for each pressure on the right-hand side.

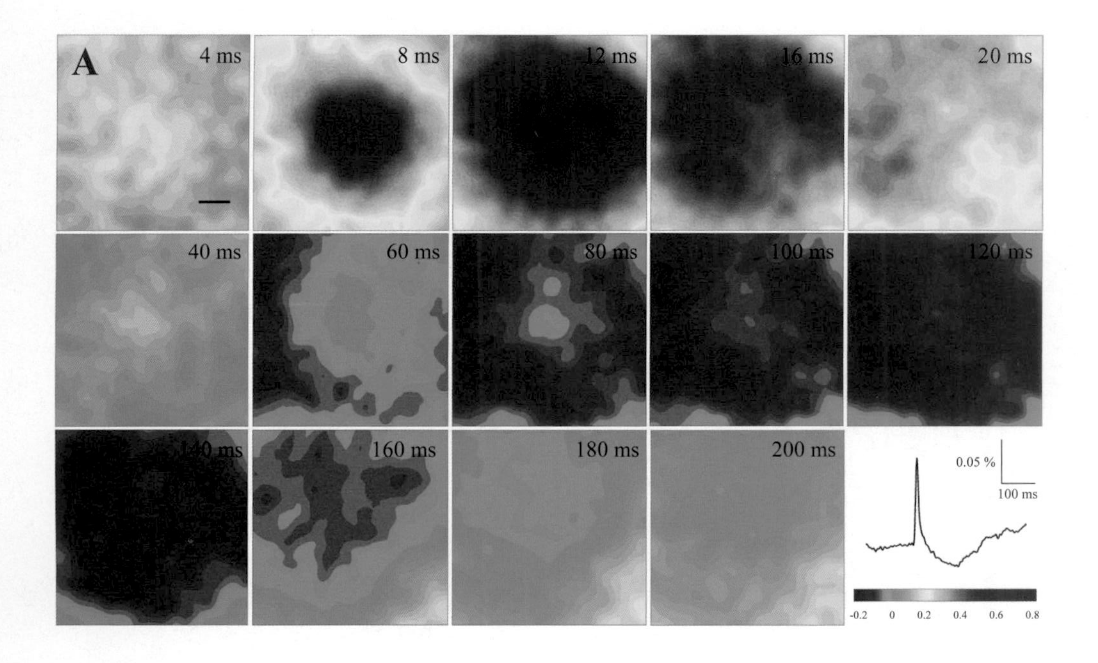
A
4 ms
8 ms
12 ms
16 ms
20 ms
40 ms
60 ms
80 ms
100 ms
120 ms
140 ms
160 ms
180 ms
200 ms
0.05 %
100 ms
-0.2
0
0.2
0.4
0.6
0.8

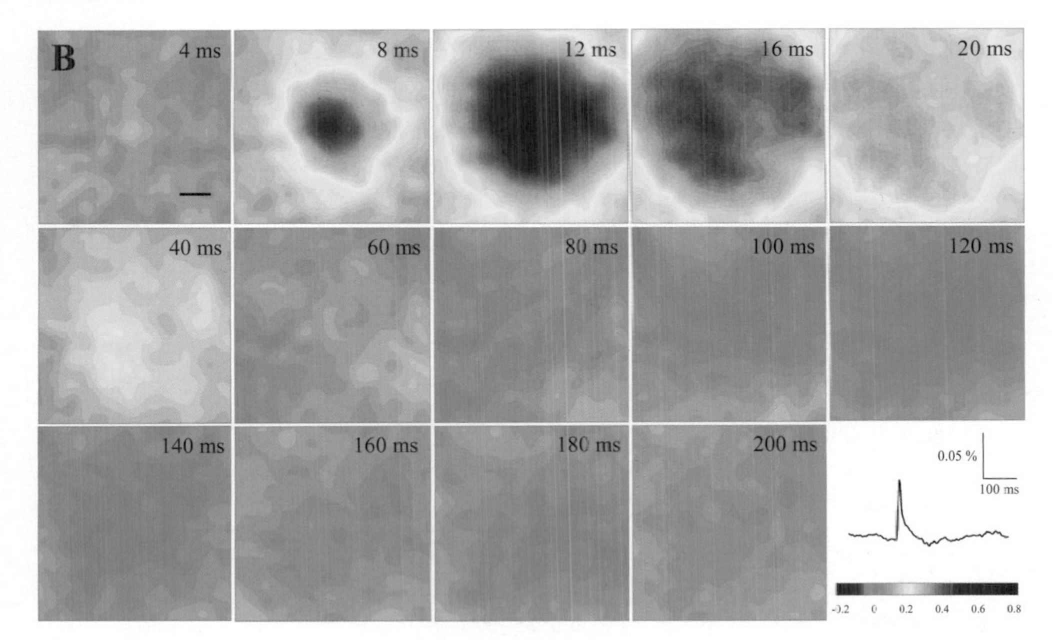

Plate 4: Assemblies evoked by whisker stimulation in the rat during light (A) and deep (B) levels of anaesthesia. Note that an inhibition (blue) occurs as a rebound after the intense activation during light anaesthesia but not after the reduced activation in deeper anaesthesia. The time after the stimulus onset is displayed in the upper-right corners; note the change in time intervals in the second and third rows. The lower-right corner of A and B show graphs of activity taken from the very centre of the assembly and charts how its amplitude changes over time. The 'rebound' is illustrated in the graph in A immediately after the amplitude of the assembly peaks and activity drops markedly below normal levels. Scale bars indicate both amplitude and time in these graphs.

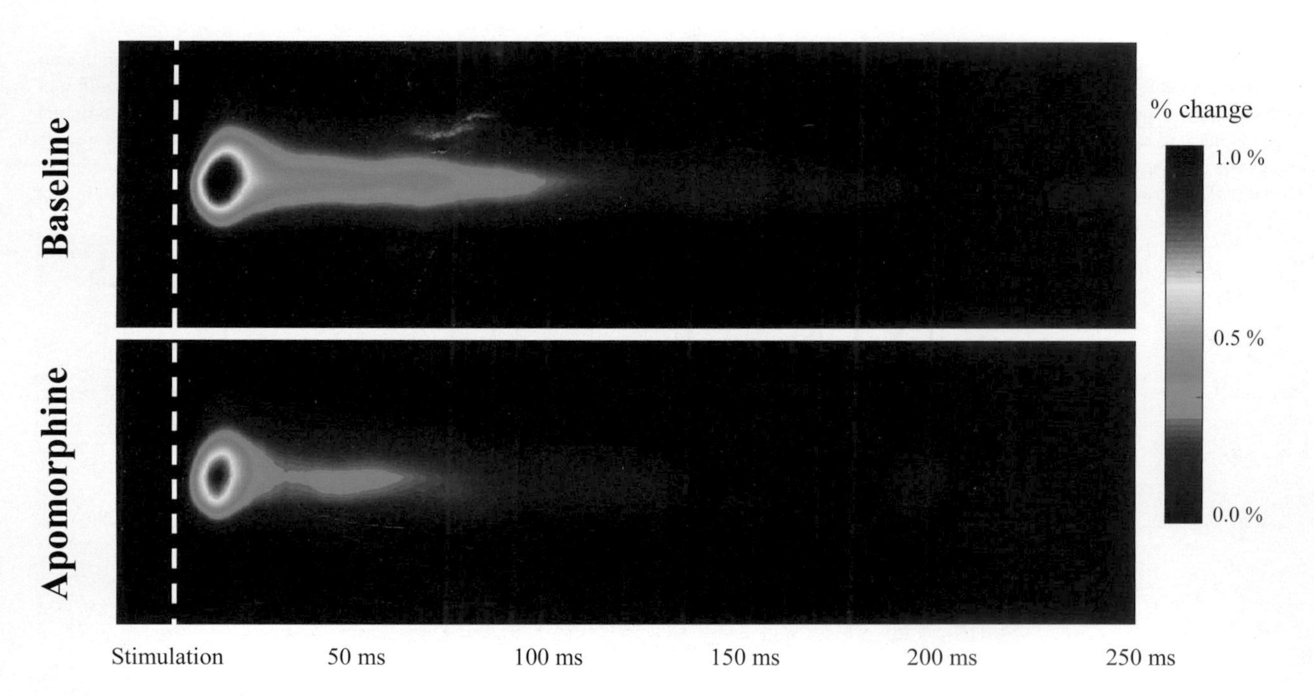

Plate 5: An assembly generated in a rat-brain slice (cortex) over time. The vertical dashed line indicates the point at which an electrical stimulus was applied to the slice. Note the reduction in intensity (less dark red and subsequent pale blue), as well as in duration, when the dopamine impostor apomorphine (lower panel) is applied (Badin and Greenfield, unpublished).

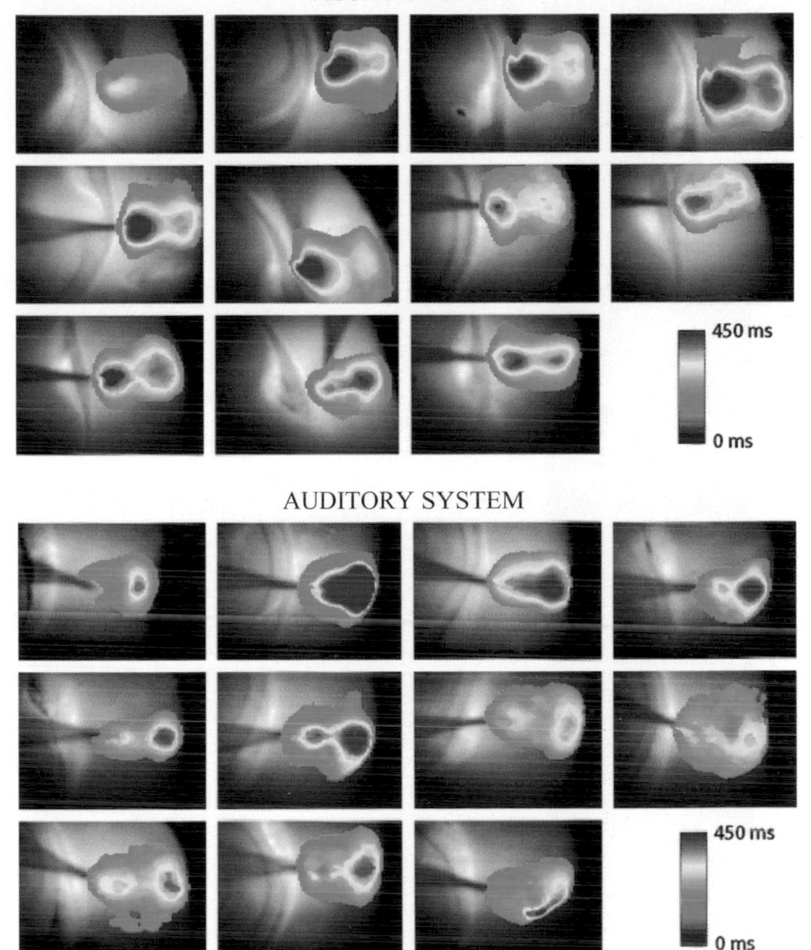

Plate 6: These two optical images show a difference in assemblies between the visual and auditory systems in eleven different experiments. Colour represents the time at which the fluorescence signal of a particular location finally decreased to beneath 20 per cent of the maximum value: 300msec. Red areas represent areas with highest activity. Note that this activity is concentrated in the deeper layers in the visual cortex and to the more superficial layers in the auditory cortex.

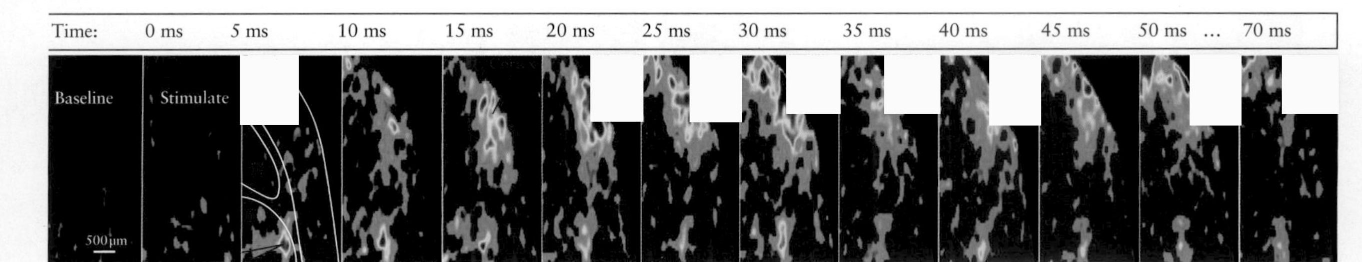

Plate 7: Neuronal assemblies in a rat-brain slice visualized with voltage-sensitive dyes (from Fermani, Badin & Greenfield, submitted for publication). While normal synaptic signalling takes 5msec to travel up to 2 millimetres from thalamus to cortex (see frame showing activity after 5ms), the ensuing assembly takes four times longer to spread some 0.5mm in radius (see frame at 20ms).

different types of *interaction* between the individual and that space, such as the potential for spontaneous meetings, the opportunity for walking, and communal versus personalized or private areas. Finally, also important are features catering for the *subjective reaction* of the individual to the space which gives them a strong sense of uniqueness, of company brand as well as an individual identity. Only when an individual feels confident enough in their own unique identity will they be able to have unique thoughts and thereby perhaps be creative.

For humans, the effects of the physical environment will be secondary to social interaction, which is in turn related to identity: your role within the group. While colours and empathetic spaces may induce a sense of well-being, they will be the background, literally, to what is happening inside your head, as the neuronal connections form and re-form ceaseless alliances (assemblies). The workplace is not a location in which it's likely that you'll blow your mind or have a sensational time: you will be self-conscious pretty much all of the time. And so it could be the perfect place for you to establish and exercise your individual identity.

As such, the stone triggering consciousness at any one moment will not be forcefully thrown, nor will the throws be frequent, in rapid succession. Rather, the effect of the throwing of the stone will be due to its large size, the many neuronal connections that, in turn, are processing your self-conscious behaviour, the reactions of others to it and your response to these. Your arousal level will not be overly heightened but, at the same time, you will not be sleepy: the modulating chemicals will be playing their part effectively enough, such that the large stone, with no competition from some new successor, and even though thrown gently, will be able to generate extensive ripples: your consciousness will be deepened. The balance between a mild sensory input and inner cognitive processes will, in an ideal world, enable you, if not necessarily to have a new idea, then at least to follow a productive train of thought . . . just like the one you're having now.

How long have you been lost in your thoughts? You look around at the blank, bland walls, the window facing on to the car park, the innocuous beige carpet and, finally, at the computer passively standing by, its screen asking nothing from you. But you haven't been creative, let alone even very productive today: surely this is not who you are – someone who simply reacts but never initiates. Would spending the day walking around in a different type of space have led you to have been more effective and happier? Perhaps. Your eyes slide back to the statutory, framed snapshot of your family: your wife, son and mother-in-law frozen in time, trying to look cheerful but somehow forced and failing. Perhaps they'll be waiting for you to return, but then again perhaps not. In any event, you might as well go home.

6

Problems at Home

It is the end of another day . . . but not really: an evening with the family now beckons. Your wife, son and mother-in-law will be waiting for you back home, and yet you contemplate returning with mixed feelings. To be honest, you're not exactly looking forward to heading back. Things have not been going well at home . . .

For some time now, your fourteen-year-old son, Jack, has been behaving out of character: he seems to have changed completely from the happy, curious, affectionate boy he used to be. Nowadays, when he returns from school he retreats into his room the first moment he can. When he is obliged to be with the family – say, at mealtimes – he seems oblivious to the conversation swirling around him and spends most of the time absorbed in his mobile phone. When he comes home after being out, you are worried you can now often smell the whiff of cigarette smoke on his clothes and, given his frequent absences, you suspect he might also be drinking, even taking drugs. Why has he, as with many of his friends, become so difficult?

ADOLESCENCE

Adolescence has always been a time of massive psychological and biological upheaval. Personal relationships take on a greater significance and the teenager starts to crave fun and exciting experiences. These years typically feature a profile of increased socializing, novelty and attention-seeking, risk-taking, emotional instability, impulsivity – and recklessness.[1] An interesting theory about this adolescent attraction to risky behaviours is the 'fuzzy trace theory', which holds that, contrary to accepted wisdom, adolescents don't actually have any deficiencies in their cognitive processes and, in fact, engage in a more accurate risk-benefit analysis than adults.[2] The crucial difference is that adults tend to minimize the absolute, overall risk involved, no matter what the immediate, short-term benefit. The more mature individual can see the bigger picture and take a longer view, rather than just being reactive to the immediate here-and-now situation and instant rewards.

An adolescent's propensity to be more miserable and moody may also drive them to seek out stimuli of greater immediacy and sensation. Compared with other age groups, teenagers place a much greater premium on pleasure; insight, or lack of it, into their emotions may also play a role in how/when/why they have difficulties in controlling their behaviour. In one test which gives participants a gambling task, for example, it was only the older adolescents who were able to improve their performance over time, in the same way that adults can. This shift, which occurs at the same time as the appearance of more seemingly mature, toned-down emotional changes is associated in turn with reduced levels of arousal.[3] Brain mechanisms for impulse control gradually take hold only as time advances.

These transitions in cognitive and emotional processing in the teenage years are likely to reflect gradually greater functioning in an area we encountered briefly back in Chapter 3 – the prefrontal cortex.[4] As its name suggests, this sector of the cortex sits at the front of the brain and constitutes a significant proportion of brain territory. The prefrontal cortex is a specific example of the fact that stages in human evolution (phylogeny) are recapitulated in general in those of individual development (ontogeny). Accordingly, this valuable neuronal real estate is dominant only in primates (occupying 33 per cent of the human brain compared to only 17 per cent in chimps)[5] and only delivers its full contribution to brain processing in humans in the late teenage years or early twenties.[6]

Moreover, since it functions through extensive networking with all other types of brain regions[7] – arguably, more than any other area – the prefrontal cortex enhances brain cohesiveness. As the teenager matures, the connections between their frontal lobe and deeper brain areas increase, a process linked directly, by their early twenties, to a growing control over performance. A particularly telling feature specific to the adolescent brain is that widespread activity is commonly detected in imaging studies that is unrelated to any particular task performance, almost as though random and disconnected thoughts are just flitting through the mind. Such spontaneous and undirected goings-on wane as adulthood is reached, suggesting that, as we mature, our brains operate a more organized collection of networks, resulting in more efficient processing. Research does indeed show that resting-state activity shifts during development into a more integrated network-activity pattern, extending out to more distant brain areas; long-range, synchronous activity across regions that are all activated collectively also increases with maturity. This transition, as the teenage years draw to a close, paints a picture of improved communication between different brain regions as adulthood progresses.[8]

Yet not everything comes of age simultaneously. Different parts of the brain mature at a differential rate, which could, in turn, explain typical adolescent social development but a persistent lack of behavioural restraint. Brain regions deep below the cortex are fully operational significantly earlier than the more sophisticated processing centres such as the prefrontal cortex.[9] Consequently, there is an uneven balance, tilting behaviour more towards immediate reward, and risk: the still-underactive prefrontal cortex will not apply the precautionary brakes to the immediate, reactive 'Go!' signal firing off in more primitive areas. Moreover, it is at just this stage that there's the highest ever lifetime peak in brain dopamine activity,[10] which will inhibit the already underactive prefrontal cortex still more.[11]

Another feature of the chemical landscape of the teenage brain is a surge in oxytocin, a hormone closely related to emotional bonding, also known as the 'feel-good factor'.[12] This neurochemical cocktail of dopamine and oxytocin leads to an increase in sensation-seeking and risk-taking in adolescence.[13] The pull of excessive dopamine and oxytocin, combined with the push of an underactive prefrontal cortex, anchors teenagers much more readily in the here and now, a state of play that trumps cognition and reflection. The overall scenario is one characterized by the sensational press of the moment, as opposed to an embedding of that moment in a 'meaningful' past-present-future narrative.

Sometimes, this state of affairs can persist into adulthood, namely, in schizophrenic patients, who appear never to lose certain aspects of the immature brain. Schizophrenia is a complex and sophisticated mental condition involving a dramatic departure from 'reality' as it would be perceived by most people and is hence regarded as a psychosis. This book is not the place to venture into an in-depth exploration of this fascinating condition, but it is worth exploring it briefly. The key features of schizophrenia include a short attention span, disordered logic, delusions, inability to interpret proverbs, unusual speech patterns and excessive distractibility.[14] Schizophrenics place a higher emphasis on any intrusions from the outside sensory world, compared to 'normal' adults, and can even feel that, unbidden, the external environment is imploding in on them and that other people can see and hear their thoughts: not for the schizophrenic that reassuring, familiar feeling of a firewall between their brains – or rather their minds – and the incoming wildfire of sensory stimulation encroaching on it. If it is indeed the case that, during development into adulthood, the sensory world is overtaken by a more 'cognitive' one then, in schizophrenia, this transition is far less emphatic. The senses remain overly dominant, while personalized associations, interpretations of the world, 'meaning', are far more fragile and idiosyncratic.

Both children and schizophrenics are easily driven by events in the immediate, external environment: as any parent knows, a child who cries over an ice-cream dropped from their pushchair can immediately be made to smile by being urged to look up at a birdy or a plane flying overhead. Meanwhile, in schizophrenia, brain responses reveal that this characteristic vulnerability to distractions has continued beyond childhood. In one experiment, members of a group of patients, alongside a healthy control group, were instructed to ignore background auditory stimuli while they performed a simple cognitive task of classifying capital letters and digits; compared to the controls, the schizophrenics showed greater levels of distraction and so took longer to complete the task.[15] If the external world is excessively demanding, it's perhaps not surprising that the young and schizophrenics alike are deemed to have a short attention span.[16]

Another telling parallel is that neither children nor schizophrenics can interpret proverbs: if asked to explain the statement 'People who live in glass houses shouldn't throw stones', a typical answer from someone suffering from schizophrenia might be something like, 'If you live in a glass house and I throw a stone at it, then your house will break.' Similarly, a child also takes the world literally, at its sensory face value. A little boy or girl admonished not to 'cry over spilt milk' might look around in surprise and incomprehension at the absence of an overturned glass. The central issue is that schizophrenics and children both have trouble understanding one thing in terms of something else. While, for children, the full complement of brain-cell connections that would underpin the requisite cognitive conceptual framework simply doesn't as yet exist, for schizophrenics, the same end result could occur for different reasons. One of the characteristic features of schizophrenics[17] is that, among many other discrepancies in the brain, there is a functional excess of the modulating, fountaining chemical dopamine.[18] So, one contributing factor could be that the effectively high levels of dopamine will suppress brain connectivity, particularly in the prefrontal cortex, where dopamine is inhibitory,[19] thereby reducing the robustness of internal cognitive processes and thus, in turn, emphasizing disproportionately the impact of the incoming senses.

Another tendency which is most probably linked to their underactive prefrontal cortex, which, as we've just seen, is only fully mature in the late teenage years,[20] is that both children and schizophrenics are more reckless than the average adult.[21] Neuroscientists have known for a long time that an underactive prefrontal cortex is linked to a greater tendency to take risks: back in the nineteenth century, one Phineas Gage, a railway worker, was tamping down explosive with a thick rod when a premature explosion drove the formidable object through his prefrontal cortex. The result, which has become the prototype case of a now established 'frontal' syndrome, was that Phineas became quick to anger, more emotional and child-like – and more reckless.

Both childhood and schizophrenia, then, conjure up a sense of living in the moment and a momentary reactivity to the stimuli incoming from the raw senses, rather than a more proactive, thoughtful, 'cognitive' take on life. The states of both the immature and the schizophrenic brain would match up with an underactive prefrontal cortex caused, in children, by the late maturation of that area and, correspondingly, in schizophrenics by an excessive amount of functional dopamine, which characterizes that disorder[22] and which is inhibitory on the cell populations in that region.[23]

Interestingly enough, there is also a third, utterly unexpected group of people with an underactive prefrontal cortex, who are also characteristically reckless:[24] those with a high body mass index (BMI) and thus heavy relative to their height – the obese.[25] Here, as with schizophrenia, and in childhood, the sensory present could be trumping the long-term consequences: for the significantly overweight, the press of the here-and-now environment is once again unusually paramount. After all, anyone biting into a cream cake or other high-fat or high-sugar treat is well aware of the consequences to the waistline: yet the immediate pleasure of the taste wins out over all else. Finally, let's look at compulsive gamblers: once again prefrontal activity is reduced in those who live for the thrill of the moment, say, as the roulette wheel spins, the dice is thrown or the chosen racehorse just starts to overtake another.[26]

So if, in all these cases, an underactive prefrontal cortex parallels a drive for sensation regardless of consequences, it could mean that in contrast, a normally active, fully matured prefrontal cortex will somehow link the past and future with the present and provide a bigger cognitive picture. One further clue pointing to this idea is that damage to the prefrontal cortex can lead to 'source amnesia', in which the memory is still intact but it is more generic and removed from a specific context or episode.[27] The patient is not in lock-step with a continuous narrative of particular events but is rather adrift in an ill-defined, hazy present informed by generic memories.

But does this mean that the prefrontal cortex is some kind of control switch for freeing us up from the press of the moment? The short answer is: no. We should think of this brain region not so much as a boss but as a facilitator or, better still, as an arch-networker. We've seen just now that a crucial feature of the prefrontal cortex is that it has more inputs to all other cortical areas than any other region of the cortex and therefore plays a key role in operational brain cohesion.[28] So, if this critical communications exchange is underactive, for whatever reason, then there will be a dramatic reduction in holistic brain operations, which are normally coordinated in order to access memories and plan ahead. How does this set-up help us with insights into consciousness?

To recap on the stone-into-the-puddle metaphor: the size of the stone will be the degree of hard-wired neuronal connectivity (the cognitive factor); the force with which it is thrown will be the intensity of the stimulus triggering that moment of consciousness (the sensory factor); the viscosity of the puddle water will be the availability and concentration of various modulators; the frequency with which subsequent stones are thrown, and their respective size and the force with which they are thrown, will determine the extent of ripples on any one occasion, namely, the turnover, and hence the size, of an assembly, the degree of consciousness at any one moment.

So, it is possible to use the stone-into-the-puddle model to predict a basic alternative mode of consciousness to that of the normal adult human – one characterized by features common to childhood, schizophrenia, being obese or compulsively gambling. Strong, raw, sensory – and thus 'meaningless' – stimulation will not depend on idiosyncratic personalized connectivity: the stone will be small. The force with which it is thrown will be disproportionately strong: the loudness of the music, the brightness of the colours, the powerful smell of the food, the speed of the spin of the roulette wheel. Next, the competition from subsequent stones will be high: the fast-paced environment will mean that no nascent assembly can extend to its full potential. Finally, this competition will be enhanced by excessive dopamine (changing the viscosity of the puddle), which will enable the over-easy recruitment of new assemblies, as well as inhibiting the prefrontal cortex, thereby fragmenting macro-level brain connectivity. This alternative state will be therefore one of unusually small assemblies, correlating with a phenomenology where the press of the senses is far more powerful than otherwise; this set-up comes at the expense of the cognitive. These two basic modes are summarized below, in Table 1.[29]

'Mindlessness'	'Meaningful'
Sensation	Cognition
Strong feelings dominate	Thinking dominates
Here-and-now	Past-present-future
Driven by external environment	Driven by internal perceptions
Little 'meaning'	Personalised 'meaning'
Not self-conscious	Robust sense of self
No time or space frame of reference	Clear episodes that are sequentially linked
Children, schizophrenics, gamblers, high BMI, drug-takers, recreations (e.g. fast-paced sports, sex, dancing)	Normal adult life
High dopamine	Less dopamine
Prefrontal under-function	Normal prefrontal activity
World meaningless	World meaningful
'Small assembly correlate of consciousness'	'Larger assembly' correlate of consciousness'

While it has been ever thus, the erstwhile balance between the mindless and the meaningful now seems to be becoming more lopsided. Our current environment of screen domination, especially for the young 'digital native' (arguably, defined as anyone born after about 1990), is increasingly characterized by two dimensions and two senses only: hearing and vision. One recent survey showed that in the age group from thirteen to eighteen years of age, daily media usage was eleven and a half hours a day – but, even more incredible, this figure jumped to over eighteen hours when multitasking was taken into account.[30] It follows that, if the environment is being transformed for so much of the time into a fast-paced and highly interactive two-dimensional space in such an unprecedented way, the brain will adapt correspondingly, be it for good or ill, also in unprecedented ways.[31]

Most significantly of all for present purposes, however, is that evidence is accumulating that videogames,[32] as well as social networking sites,[33] enhance release of dopamine in the brain. This neurological profile is similar to that seen in drug addicts and is associated with abnormalities of the prefrontal cortex.[34] It is as though screen activities, so popular particularly with adolescents, are tilting and exaggerating even further the age-old balance between mindless and meaningful. On the one hand, the excitement, compulsion and sheer pleasure of the game enhance brain dopamine levels, while the screen displays appeal to the sensational rather than the cognitive mode. This drive for fast-paced, literal stimulation could make it more likely for the small-assembly mode to be the default setting in the consciousness of the digital native, to an extent it never has been in previous generations.[35] No wonder Jack is as he is . . .

DEPRESSION

If Jack is too much submerged within the external world, albeit an online one then your wife, Jane, has the opposite problem: how you long for her to be *more* engaged with the outside world. For some reason, she has not been coping well lately. She seems listless and uninterested in anything you suggest that might cheer her up. She no longer goes to the gym and, although she spends much of her time these days in bed, she seems tired all the time. She can't be bothered, most of the time, to eat a proper meal, or wash and dress, and dissolves into tears at the slightest provocation. Everything seems too much effort for her, and sometimes she just curls up under the duvet, quietly crying to herself, seemingly for no reason. Nothing makes her laugh any more.

One of the symptoms of this profile – clinical depression – is anhedonia: literally, a lack of pleasure. The depressed individual feels numb and indifferent to, say, the beauty of a sunset. They feel overwhelmed by everything and unable to cope: they withdraw from the complexity and demands of daily life. In turn, the outside world seems remote, grey and muted. Whether this under-stimulation of the senses is because the outside world already feels distant or, conversely, whether some under-performing sensory system somehow feeds into the feeling of being cut off and isolated is a chicken-and-egg question that bedevils brain research in general. In any case, the two scenarios need not be mutually exclusive. Either way, it means that locked away in their own inner world impervious to, and undistracted by, the press of the sensational, the depressive is able to rehearse over and over again a strong continuity of thought, perhaps even an obstinacy of the same, persistent imagining.

In depressed individuals, there is a shift in the whole chemistry of the brain, an imbalance, yet again, of the modulating chemical family of amines – dopamine, noradrenaline and serotonin – which underpin arousal levels and account for the metaphorical viscosity of the puddle. Accordingly, depressed patients are likely to be prescribed something like Prozac, a medication described generically as a 'selective serotonin re-uptake inhibitor' (SSRI), that is, a drug that enhances the availability of (primarily) serotonin. It's not immediately obvious why an increase in serotonin, along with its chemical cousins dopamine and noradrenaline, should alleviate lowness in mood, but the story of anti-depressant treatment has its roots, as does much in scientific discovery, in early observations that occurred quite by chance . . .

In the 1950s, the prescribed treatment for high blood pressure was a drug called reserpine.[36] The rationale was, presumably, that if noradrenaline and its close relation adrenaline activated the 'fight and flight' mechanism of the body by increasing heart rate and raising blood pressure, then a drug that reduces the availability of noradrenaline should have a therapeutic effect in reversing these effects in hypertensive patients. Reserpine does indeed efficiently deplete the body stocks of noradrenaline and, consequently, did have the desired and appropriate result of reducing blood pressure and heart rate. However, a terrible side effect was soon apparent: the admittedly now calm patients had become severely depressed, bordering on suicidal.[37]

Meanwhile, during that era, in the middle of the twentieth century, one of the most common and life-threatening conditions was tuberculosis. The treatment at the time was a drug called iproniazid, which proved effective in inhibiting a particularly virulent strain of the disease.[38] Iproniazid increases the availability of dopamine, noradrenaline and serotonin by virtue of its biochemical action in blocking the enzyme (monoamine oxidase) that would normally degrade these transmitters: it's known therefore as a monoamine oxidase inhibitor (MAOI). However, once again, there was an unforeseen side effect: the TB patients became surprisingly euphoric, not just because they were recovering from a serious illness but directly as an effect of the drug.[39]

These two serendipitous findings, when taken together, suggested a fascinating possibility. If a drug such as reserpine, which depletes the body of the amine chemicals, makes people depressed, and if another drug, like iproniazid, which enhances the availability of those same amines, makes people euphoric then, clearly, mood could in some way be linked to the amine family. The Amine Theory of Depression is, accordingly, that a key feature in determining mood (affect) is the functional availability of amines in the brain: if levels are high, so will be the ensuing mood. However, if the levels are abnormally low, clinical depression is the phenomenological corollary.[40]

However, there is more to depression than the turning on or off of an amine tap. For example, we know that there's a 'therapeutic time lag' of some ten days from initially taking the drug to seeing any beneficial effect.[41] So it's not simply a question of raising serotonin levels but of understanding more about the long-term effects of this increased serotonin, and other amines, on the relevant neurons. One idea is that the target receptors for serotonin will need time to recalibrate their response to the newly elevated levels of the modulator: another, not mutually exclusive, possibility is that the drug-induced boost in amine levels triggers a separate, longer-term reparative brain mechanism such as the release of slow-acting growth factors and, possibly, a reduction of inflammation. So not only will 'bottom-up' chemical and synaptic factors come into play but more cognitive 'top-down' ones play a role, too. It may not be surprising that more protracted, longer-lasting neuronal plasticity will operate in the depressed brain by adapting to the new chemical landscape which has been induced by the therapeutic intervention.

However, the natural plasticity of the brain can also be harnessed to modify mood without any external drugs. For example, the psychotherapeutic approach of Cognitive Behavioural Therapy (CBT) for depression has proved particularly successful in recent years.[42] As the name implies, no medication is used in CBT; instead, the therapist helps the depressed patient see the world from a different perspective through dialogue. One way of interpreting CBT in neuroscientific terms would be to regard it as a particularly effective form of contrived plasticity at work, where neuronal connections are being rearranged in such a way as to help the patient see the world differently: a new stone would be formed.

The more we interact with the environment, the more brain-cell connections will change: however, it's been recognized for decades that depressed people withdraw from the world, shunning interaction with it and, in particular, avoiding physical exercise. Therefore, they are likely to have less reason for a raised heart rate, and hence their levels of amines such as noradrenaline which accompany high arousal will not be as high as otherwise. A fascinating recent discovery is that depressed individuals also undergo a decline in production of new brain cells – the neurogenesis that we saw previously (Chapter 3) which was enhanced by environmental enrichment and voluntary physical exercise and which facilitates an ever greater plasticity.[43] A virtuous cycle opens up of being receptive to new stimuli and experiences, experiencing greater arousal, raised levels of amines and, as a result, heightened mood. The opposite scenario would be a decline in neurogenesis accompanied by a disinterest in the outside world, less excitement and thus a depressed mood: a vicious cycle this time, in which, as with the interplay of exercise and well-being, and the interplay of under-stimulation and feelings of disengagement, it would be very hard to determine whether the chicken or the egg came first.

What might happen to assemblies in the depressed brain? Imagine an individual withdraws from everyday activities, let alone strenuous exercise, who spends most of the day under the duvet, or at least less time interacting and engaging with the outside world generally: they will have fewer experiences and their neuronal connectivity is far less likely to change, simply because much less will be going on, with little new stimulation coming into the brain. Existing connections will persist and thereby become stronger with continued use, that is without challenge. Moreover, the brain will not have appropriate levels of the modulators dopamine and its amine siblings to aid and abet the ready formation of new assemblies and increase their turnover. This means that there will be an ever greater chance of the same old, abnormally large stone driving pond ripples – an eventual assembly – unopposed by any rival: the slower the turnover, the more extensive the assembly could be. If, on top of that, the depressed focus is persistent and ruminative, then ever more connections will form and the focal stone will be ever larger, with a still more extensive assembly. So, could clinical depression consequently be correlated, this time, with unusually large assemblies?

If so, then the very factors that determine the small assemblies of the 'mindless' brain state would now be the polar opposite. And such seems to be the case. For example, the small-assembly 'mindless' state, which was previously characterized by an underactive prefrontal cortex, is in clear contrast to the situation dominant in depression, where the prefrontal cortex is over-active.[44] Also, the abnormally small assembly that featured in sensational experiences was generated with high levels of dopamine an amine which, along with its chemical relatives, as we've just seen, is deficient in depression. Finally, part of the reason for the small-assembly brain state would be that there was rapid competition from new assemblies waiting to form as a result of intense interaction with the outside world – the very antithesis of the depressed condition, in which the individual shuns interaction with external realities.

If unusually large assemblies correlate with depression, this would help explain the still-puzzling efficacy of electroconvulsive shock therapy (ECT), in which the severely depressed patient has an electric current passed directly through their brain via surface electrodes.[45] ECT, though on the face of it crude and non-specific, is nonetheless often effective when all other drug treatments have failed. Perhaps what is happening is that the electrical current is disrupting the obstinate neuronal networking – indeed, a common side effect of ECT is loss of memory – thereby facilitating the formation of new and different connections: a kind of enforced plasticity. By smashing up the large, original, persistent stone, so new, more normal-sized stones can be formed more readily.

The idea that depression could be correlated with abnormally large assemblies would also fit well with another clinical riddle: the success of the drug lithium in treating bipolar disorder, in which the patient experiences exaggerated mood swings, from deep depression (let's say, large assemblies) to hyper-excitability, or mania, which would have many of the features profiled for the small-assembly mode.[46] One possible explanation could come by the fact that lithium is a chemical neighbour to sodium in the periodic table of the elements: this means that it can act more readily as a chemical impostor in the brain, substituting for sodium ions. The very basis of communication between neurons is the influx of sodium ions into the neuron which causes the initial trigger for the all-important electrical blip, the action potential, whereby the cell is active. As it happens, lithium can compete with sodium to pass through its special channel into the interior of the neuron[47] but, once there, it will be unable to complete the job: the generation of action potentials will now be compromised.

Why should such a generic nuts-and-bolts action be so effective in suppressing a sophisticated and complex cognitive problem such as mania? Manic depression is appropriately known as bipolar disorder, because of the two extreme polarities of mood between which the patient swings. The mania phase is the complete antithesis of the numbness and inertia of depression: instead, it is characterized by hyperactivity and impulsive behaviour, a mindset trapped in the present, overly reactive to and distracted by the outside world – in short, the small-assembly mode attributed here to the child or the schizophrenic. In line with this suggestion, bipolar patients are more reckless and also have an under-functioning prefrontal cortex,[48] as in schizophrenia.[49] So, if the manic phase of bipolar disorder can be interpreted in terms of unusually small assemblies, and if the core action of lithium is to impede the basic machinery of neuronal activity, then perhaps the efficacy of lithium will be to impede the over-easy formation of one assembly after another, the high turnover that could characterize the mindset of the child or schizophrenic, and the bipolar patient in the manic phase. In short, lithium would act to *stabilize* the generation of assemblies by applying a chemical handbrake.

But a nagging question remains, in that this handbrake applies only to mania: lithium is ineffective in the depressive phase of bipolar disorder, and in standard depression, too. Why? The crucial clue could lie in the fact that lithium turns out to be effective in depression – so long as it is given *before* the negative mood takes hold.[50]

Although it's just a hypothesis, one possibility is that the key to the action of lithium is steadying neuronal activity to a baseline norm: hence its lack of any effect in healthy individuals for whom assemblies are already stable, and indeed in depressed patients for whom an assembly has already grown unusually large and long-lasting and which is therefore already established. In contrast, while assemblies are forming, as in mania, or indeed about to extend, in the beginning of the depressed phase, that is exactly when neuronal propagation – namely, raised activity – may be most critical and most effectively targeted by lithium.

PAIN

Another similarity between schizophrenia and mania, which, again, is the complete opposite to the profile of clinical depression, is sensitivity to pain. While schizophrenic[51] and manic individuals[52] have a raised threshold to pain – namely, they feel it less – depressed patients have a lower threshold for pain than normal: they feel pain to a greater degree, a condition known as hyperalgesia.[53] So, if depression is correlated with unusually large assemblies, and if this brain state, in turn, is linked to greater pain perception, then could large assemblies more generally be linked to the subjective degree of pain for all of us? There are several reasons why this intriguing possibility might be the case.

One is that pain is expressed as other associations – stabbing or burning, for example – and these associations depend on extensive neuronal connectivity.[54] There is also a daily threshold to pain: for example, in a project on healthy volunteers, the experimenters applied electrical shocks to the subjects' teeth and established that the best time to have a painful experience is in the middle of the day.[55] This fluctuation in pain threshold could be due to the modulating fountains of amines, which have a daily rhythm. A further aspect of pain perception is that the more it is anticipated, the greater it seems,[56] presumably because the extended time frame gives the chance for an assembly to build up.

Yet another reason for suggesting abnormally large assemblies might be at work relates to 'phantom limb pain',[57] in which amputees still feel sensation in a missing limb. Physiologists Ronald Melzack and Patrick Wall developed long ago the notion of the 'pain matrix' – arguably, comparable to an assembly – whereby, if you don't receive the correct feedback from your limb (because it is no longer there), neurons will be overly stimulated and a larger assembly produced: hence a greater sensation of pain.[58] In addition, the drug morphine, a powerful painkiller, results in a 'dream-like' euphoria: when patients take it, they say they feel the pain but it no longer matters to them.[59] We know that morphine acts as a blocker at target receptors for a naturally occurring opiate (enkephalin) that tends to have an inhibitory action in the brain and which would thereby reduce the size of the assembly.[60] And, finally, schizophrenics – with a suggested predominantly small-assembly state – have a very high threshold for pain.[61]

Moreover, we saw in Chapter 2 that, typically, anaesthetics defy a general, unitary explanation due to the diversity in their chemical structure and their bottom-up cellular actions: but, crucially, what they could have in common as their key action is that, at the mid-level of brain operations, they all reduce assembly size. This would mean, as the anaesthetic was taking its effect, that there could be an interesting and counterintuitive paradox. If the anaesthesia comes on slowly, as it reduces your capacity to generate only small assemblies, then it would mean that *before* you lost consciousness you would experience a feeling of euphoria characteristic of the highly emotionally coloured small-assembly state. This is certainly a strange idea, but it is one that seems to hold true: in the old days, before anaesthetics were as efficient as they are now, patients about to undergo surgery went through stages of mania and excitability and, indeed, people would indulge in 'ether frolics'[62] or take nitrous oxide as a recreational drug[63] – just as they are starting to do once again;[64] moreover, the drug ketamine, an anaesthetic, is frequently used in low doses as a recreational drug.[65]

By comparing the possible subjective phenomenology of large and small assemblies, a more general principle emerges. Bear in mind that the depressed patient reports that they are emotionally 'numb': it is not so much that they are feeling very sad but that they are feeling *nothing at all*. Meanwhile, the child, the manic individual, the drug taker and the schizophrenic are all experiencing strong emotions, albeit both positive and negative. These clues logically lead to the following hypotheses: firstly, the more extensive the neuronal assembly, the less raw emotion; secondly, therefore, emotions must be the most basic form of consciousness. You only have to think of the dog wagging his tail, or the baby gurgling, to find this second proposition hardly surprising.

Looking through the prism of neuroscience, we can now interpret clinical depression as an unusually large assembly: the stone would be large, as a wide and highly personalized neuronal connectivity would be operative. In addition, this enhanced connectivity throughout the brain would be favoured by an overactive prefrontal cortex: there would be no competition or other distractions from the outside world, no other stones would be forcefully thrown and, indeed, the senses would be muted – the outside world un-sensational. A very sluggish turnover of assemblies could then grow unopposed. Arousal levels would be low and, as a result, amine levels would also be low, and the viscosity would be thick, further preventing any ready formation of rivals. And, just as inactivity and lack of interaction with the outside world (a lack of stones being thrown) is a sign of depression, so depression can result from inactivity and disengagement.

In the elderly, depression and lack of physical interaction with the outside world can be particularly marked.[66] In a survey comparing cognitive function in community-dwelling and institutionalized old people of normal intelligence, researchers found that older individuals living in a community dwelling performed better in cognitive tests than those that were institutionalised.[67] One way to test the effects of different types of lifestyle in older, healthy adults is by looking at the relationship between their lifestyle and their 'cognitive reserve', defined as the degree to which the brain can create and use conceptual networks as the brain ages. Perhaps unsurprisingly, a greater involvement in intellectual and social activities is related to less cognitive impairment, which substantiates the 'use it or lose it' mantra. A mentally active lifestyle could protect against mental decline by increasing the density of synapses between cells in the brain, thereby improving the efficacy of signalling from intact neurons.[68]

Memory training has a proven benefit in the brains of the elderly.

In one study, researchers conducted brain scans on participants before and after an intensive eight-week memory-training programme.[69] They found that performance improved and cortical thickness increased in the group which had undergone the memory training, thereby providing further, and very reassuring, evidence that the capacity of the brain to undergo plasticity continues into old age. The more experiences leaving their mark on the brain through its continued adaptability, the greater the 'understanding' – the ability to see one thing in terms of many others – which reflects the wisdom we associate with older age.

Now, however, imagine the scenario envisaged back in Chapter 2 of a mature, carefully crafted individual brain with connections that have responded to, and been activated, strengthened and shaped by, sequences of specific experiences that no one else has ever had, nor ever will have: a unique individual mind. Next, imagine that those highly individualized connections are being slowly dismantled as the branches (dendrites) wither away. The potential stones available for throwing into the puddle would be reduced to mere pebbles. An individual would become like a child again, in that they would no longer have at their disposal the checks and balances of the adult mind against which to evaluate ongoing experiences: people and objects would no longer have the highly personalized significance so carefully accumulated over a lifetime. We would, in this case, see the sad and tragic symptoms of Alzheimer's disease, a form of dementia, in which the patient is, literally, 'out of their mind'.

DEMENTIA

Daisy, your mother-in-law, now lives with you and, tragically, she is in the early stages of dementia. The symptoms, as you are all too well aware, are as wide-ranging as they are devastating: you've been able to group them into three main categories. You can list the *cognitive* symptoms as difficulties with memory, a reduced ability to learn new things and an impaired understanding and problem-solving ability. Daisy is starting to have a poor recall of names and places and, once or twice, has wandered off and got lost. Occasionally, she'll mislay things or put them in strange places. She is also showing a decline in judgement, a lack of awareness of danger, or dressing inappropriately, for example going out without a raincoat in a downpour. Increasingly, she is struggling to read, write or hold a conversation, as she has trouble with her vocabulary: one instance was when she spoke of 'a lot of water coming from the sky', since she'd mislaid the word 'rain'; on another occasion she talked of a 'house with wheels' because she couldn't instantly generate the word 'caravan'.

Then there are the *emotional* symptoms, such as irritability and difficulties with social interaction. People with Alzheimer's disease become easily agitated over small things, as well as paranoid, which leads to them feeling suspicious of others and accusing them inappropriately of wrongdoing. Another sign of a loss of the normal neuronal checks and balances is a tendency to be overly familiar or sexually uninhibited in conversation. Dementia patients are often impatient with others, for example if they have to queue. The emotions they express may be out of context, such as unsolicited aggression or inappropriate laughter or tears. Finally, there are the more obvious *physical* symptoms, especially difficulties carrying out normal daily activities, such as cooking a meal, washing, dressing, gardening, and so on.

Every seven seconds, someone in the world is diagnosed with dementia. It is more common with increasing age, with only 2 per cent of victims diagnosed before the age of sixty-five. Once detected, the symptoms can be stable for up to five years, and sufferers live, on average, for a further ten years, depending on the age at diagnosis. There are 800,000 individuals in the UK currently suffering from Alzheimer's disease; this will rise to almost 2 million by the middle of this century. Worldwide, there will be a heartbreaking 80 million cases by 2040, as the global population ages. Dementia is not a natural consequence of ageing, but it is a disease that occurs in older people.[70] Generally, 70 per cent of dementia is due to Alzheimer's disease specifically, which may only be confirmed conclusively *post mortem* by certain markers which can be stained up and identified in the brain. The key feature of dementia, whether it be Alzheimer's or due to some other disease (such as fronto-temporal dementia or Lewy body dementia), is a loss of the personalized neuronal connectivity which, as we've seen, makes each person the unique individual they are.

If you are unable to access the checks and balances of your personalized neuronal connectivity, if objects and people around you no longer have a meaning then, increasingly, you have to take the world at face value. You are back in the infant's world of 'booming, buzzing confusion'. Sometimes, when waking up in hotel rooms on holiday, that realization of where you are and why you are there takes a fraction longer than it does otherwise, before your 'mind' – your personalized cognitive take on the world – kicks in. As you grow into adulthood, being able to put the here-and-now moment of consciousness into the narrative of your life and your unique life story is something you take for granted: but the patient with dementia is not so lucky. Their consciousness will be one of permanently small assemblies, doomed to become ever more diminished: indeed, Daisy seems to be developing a more reactive, child-like, here-and-now type of mental state.

If the conscious state of someone with dementia could therefore be described by an ever smaller neuronal assembly, how then might current treatments be interpreted as having their effect, in stabilizing the current size of a patient's assemblies? Present-day drugs such as Aricept temporarily help some 50–60 per cent of the people for whom they are prescribed, with the benefits lasting around six months. At best, patient's symptoms may indeed ameliorate, or at the very least stay at the same level. If a patient remains on medication for up to a year, it can slow down the rate at which memory symptoms become worse and also lead to improvements in the way people are able to carry out everyday activities such as washing, dressing and cooking.

Aricept works by increasing the availability of the transmitter acetylcholine, one of the all-important modulators in facilitating the formation of assemblies. At the nuts-and-bolts level, it blocks an enzyme that normally destroys acetylcholine, which is now dwindling in supply as the cells that make it are lost. So Aricept will now combat this state of affairs by prolonging the availability of this precious chemical. But sadly, this means that the effects are only temporary, since the drug doesn't get to the heart of the problem; ideally, it would prevent the cells that release acetylcholine from dying in the first place.

But we now have an important clue, namely, that the clusters of cells forming the hub in the primitive part of the brain which spray out the fountains of chemical modulators could be pivotal. It has long been a mystery why the remorseless cycle of progressive loss of brain cells that characterizes Alzheimer's disease is not a generic feature of the central nervous system:[71] only certain groups of cells are vulnerable, namely the hub cells which are the first to start degenerating long before symptoms become apparent. One valuable insight, well-known to clinicians, is that Alzheimer's can frequently co-occur with the other major degenerative disorder Parkinson's disease.[72] While Alzheimer's is a cognitive impairment and Parkinson's a dysfunction of movement, and while both conditions are the direct result of the death of certain groups of neurons, it might still be the case that there is a basic, single common mechanism underlying the characteristic and continuing cycle of cell death in the first place. The hub cells where cell loss in both Alzheimer's and Parkinson's starts come from a selective part of the embryo and, as a consequence, they have some fundamentally different properties to all other brain cells. It's tempting to suggest that these properties might explain why these hub cells in particular are primarily and specifically vulnerable to neurodegeneration. It turns out that the hub cells alone retain their mechanisms of developmental growth right into maturity. The processes underlying Alzheimer's disease and, indeed, Parkinson's, could therefore be an aberrantly activated form of development.[73]

So, what effect would such a process have on the brain in terms of assembly dynamics? If these key cells selectively start to die, fewer of the amine family members will be released to facilitate a normal-sized assembly: instead, it would be ever smaller, recapitulating a slow return to infancy – a tragic feature of dementia. Still, a critical factor that differentiates the dementing brain from the developing one is that, in the infant, the small assembly would be due to modest connectivity in the various target neuronal networks in the 'higher' brain regions such as the prefrontal cortex, with the amine fountains working just fine, and perhaps excessively so. In contrast, in the degenerating brain, the modulating fountains could be thought of as drying up, even though the pre-existing neuronal connectivity can get along fine for a while: it's well established that the symptoms of dementia may only appear some ten or twenty years after the degenerative process has started. In other words, while the assemblies of childhood are small because the stone is small, the assemblies in dementia are small because the ripples cannot spread easily.

Eventually, however, the assemblies become smaller still as the stones themselves diminish and as local connectivity in networks throughout the brain is dismantled. It follows that if the stone can somehow be made bigger, then the assemblies might still become a little larger, to compensate for the less efficient ripples. In phenomenological terms, that may be just what happens with different types of non-drug therapy which offer some assistance in dementia. For example, there's Reminiscence Therapy, which has proved effective in reducing the worst effects of Alzheimer's.[74] Even though a patient might have a poor memory for recent events, they may well be able to talk about their experiences in the war, say, as though they occurred yesterday. Reminiscence Therapy is based on the fact that most people with dementia have vivid memories from the past that can be accessed to improve mood, well-being and relationships with their families, carers and other professionals helping to look after them. This is because the older memories are more stable.

We know that it takes some two years for the firm laying-down of memories; one famous case of a severe epileptic (known throughout his life to researchers as 'H. M.' but as Henry Molaison after his death in 2008) demonstrated this time lag very clearly. Henry had surgery for epilepsy: the radical removal of part of the brain linked to memory processing. Not only was he amnesic as a result about events after the surgery, but for the two years preceding it as well.

By presenting to people with dementia objects such as photos, cine films and videos, scrapbooks and music – in fact, anything that may jog a memory and develop a discussion about a familiar topic – more connections are being stimulated than would otherwise be the case. In one home for the elderly in London, for example, there's an entire '1940s room', complete with Bakelite radio and magazines from that time.

Another way to ensure that the assembly is as large as possible is through the various senses: for example with music.[75] A favourite piece of music, or one with specific relevance, such as a song that was played at the person's wedding or was popular at the time of a first date, can have a remarkable impact on Alzheimer's patients. Even as the memories weaken, music can still have a powerful, beneficial effect. Music therapy uses tunes, rhythms, instruments and singing to improve a patient's sense of well-being. Music obviously has a great power to affect our thoughts and feelings; as we saw earlier, it's a basic part of being human, an equal but opposite complement to language. A dementing patient may often be receptive to music long after other mental processes such as concentration and memory have been lost, and so the power of music to evoke these feelings can be used to communicate with people even when they struggle to make sense of things at all other times.

In his book *Musicophilia*, neurologist Oliver Sacks recounted the story of eighty-year-old Bessie T., a former blues singer with Alzheimer's disease; it had left her with such severe amnesia she couldn't hold anything in her mind for more than a minute. In preparation for a talent show in the care home where she lived, she and her music therapist practised some songs together. On the day, Bessie performed beautifully and with great feeling; amazingly, she remembered all the words. Yet just a few moments later, after she had stepped away from the microphone, she couldn't recall having sung at all.[76]

As well as using vision and hearing to promote activation of more neuronal connections, keeping the stone as large as possible, there are smell and taste. If favourite foods or drinks are used to stimulate discussion about visits to special places, events or the people they used to share them with, Alzheimer patients can find it easier to remember things. Finally, the sense of touch, stimulated by items of clothing and ornaments from the past, can also be used, along with treasured jewellery or a loved one's medals. In these examples, where modulating chemicals (the viscosity of the puddle) and neuronal connectivity (the size of the stone) are sub-optimal, the feature that can still be enhanced is the force of the stone throw.[77]

What kind of assemblies are forming and disbanding in succession in your brain right now? As you turn your key in the lock, there's silence in the house. To your guilty relief, everyone has already retired to their bedroom – for different reasons: Jack will be communing intensely with his mobile, while Daisy, like the small child she is becoming, will have already been asleep for hours. Jane will be staring at the ceiling and oblivious to your arrival. Even Bobo is slumped down in his basket, muzzle laid out between two outstretched paws, exhausted. The bitter-sweetness of solitude in the spare room is your only option.

7
Dreaming

Groggily, you climb the stairs to the small, featureless room. You tug off your clothes and slide in under the duvet, then lie on your back and stare up into the shadows. In one sense, the day has flown: it seems only a moment ago that you were here in this very place, urging yourself to get up as the drowsiness slowly and grudgingly receded. But now you are slipping back in the reverse direction. Before you know it you're asleep, and dreaming . . .

The average time that different species sleep per night varies hugely, from three hours in the donkey to twenty hours in the armadillo. It seems that this variation reflects different ecological needs among various non-human species, and constitutes a pay-off, as it did for early humans, between the amount of time needed to forage and the risk from predators.[1] But within the protracted oblivion of sleep, there are pockets of time when you'll surface into an indisputable form of consciousness somehow similar to, yet also weirdly different from, that of the day: dreaming. The term 'dreaming' itself has traditionally been used interchangeably with another, more technical one, 'REM' (rapid eye movement sleep), a peculiar stage characterized by a swift flickering of the eyes behind the closed eyelids. The distinction between non-REM (NREM) and REM sleep applies across the animal kingdom.

Recently, however, neuroscientists have discovered that the dreaming state can actually sometimes occur independently of these hallmark signs. Research shows that, although subjects most frequently report dreams mainly after being specifically woken from REM, this is not necessarily always a hard-and-fast rule: when subjects woke during non-REM sleep, researchers in one study simply asked them about what had just been passing through their heads, instead of whether or not they had been dreaming.[2] Some 50 per cent of subjects woken from non-REM sleep described some form of dreaming at all stages of sleep.

A further dissociation between the subjective experience of dreaming and the specific, physical stage of REM is apparent during night-time seizures within a non-REM period, which are often characterized by nightmares. Moreover, the effects of damage in different areas of the brain also reveal that REM cannot be exclusively synonymous with dreaming, and vice versa: lesions of the primitive brainstem abolish overt signs of REM but spare dreaming, whereas lesions of the more sophisticated brain regions spare REM yet abolish dreaming.[3] To sum up: as we start here to probe into dreaming, it is important to keep in mind that, although it's sometimes dubbed 'REM sleep', this fascinating form of consciousness isn't exclusively linked to that particular sleep stage.

THE PURPOSE OF DREAMS

To the dreamer, a dream can seem at the time eerily indistinguishable from everyday consciousness yet, in retrospect, upon waking, so self-evidently different. The fragmented narrative, the improbability of much of what happens – flying, for example, or the transmogrification of one individual suddenly into another, and so on – seem ludicrous, even embarrassing, in the cold light of day. So what purpose does this spooky consciousness, removed yet somehow linked to reality, serve? In order to answer this question, a standard game plan for scientists has been to investigate what happens when humans are denied the opportunity to dream. Given that dreaming occurs in both REM and non-REM sleep stages, it's impossible selectively to deprive subjects of dreaming yet at the same time allow them to sleep naturally through all the other stages. Nonetheless, the effects of sleep deprivation in general on mental function have been documented for over half a century[4] and all the evidence points to the unsurprising conclusion that sleep in general, and perhaps dreaming in particular, is clearly important in enabling us to deal with the vicissitudes of the coming day.

For instance, research shows that thirty-six hours of sleep deprivation leads to a significant deterioration in performance on memory tasks, even when the hapless volunteer is administered substantial amounts of caffeine to combat their drowsiness.[5] Another effect of experimentally induced insomnia is that the subjects kept deliberately awake are in a heightened emotional state,[6] while imaging studies reveal that a brain area linked to emotional responses (the amygdala), will respond with 60 per cent greater activity.[7] REM sleep, in particular, appears key somehow to dissipating the strength of emotional stimuli experienced the following day to emotional stimuli viewed the previous night – and at the same time reducing corresponding activity in the amygdala.[8] Moreover, there is an interesting parallel between the over-reactive emotional dysfunction resulting from sleep deprivation and that of various psychiatric conditions such as that induced by the drug LSD, which are often characterized by poor sleep patterns.[9]

The idea that sleep in general, and dreaming in particular, fulfils some kind of consolidation function which helps the dreamer cope with the waking hours, is far from new. However, it is one thing to observe the negative effects of extreme, contrived sleep deprivation but quite another to extrapolate what might be happening in the brain during normal dreaming – or even to understand why we dream at all. Sleep expert Allan Hobson and his colleague Karl Friston approached this most basic of questions from the irrefutable assumption that dreaming offers you a private world that appears to be generated internally without the additional burden of sensory inputs from the outside teeming into the brain.[10] Then again, the information relayed by your ears, eyes, tongue, nose and fingers is not just some take-it-or-leave-it optional extra but might provide the essential first step for dreams to do their job. Hobson and Friston's idea is that dreaming is heavily dependent on the senses,[11] on their processing earlier, while you were awake: accordingly, the full impact of what the senses have to impart can only be really appreciated with the brain 'offline'. As a consequence, the dreaming brain is dubbed a 'virtual reality generator' that will enable the dreamer, once they eventually wake again, to navigate, anticipate and make the most of the real-life environment, even if only subconsciously.

At first blush, this scheme of things seems highly plausible: that, in dreams, the dreamer is trying to find explanations for unreal visual searches triggered by input from the muscles controlling the eyes, through the eponymous rapid eye movements, thereby leading to the rearranging of neuronal connections as this fantasy world is appropriately simplified. Without this periodic offline repair ('pruning'), afforded by dreams, neuronal networks will become overly complex and dysfunctional.[12] Perhaps the brain needs temporarily to block incoming sensory input to the networks it is trying to consolidate or repair – just as a plumber will turn off the water when fixing a pipe.

But why couldn't this entire process occur during ordinary old sleep, without the subjective experience of a dream? Although the earliest ideas on the function of sleep more generally were based on the scenario of some kind of preparation, be it of the restocking of brain chemicals or the rehearsal of appropriate reactions, the awkward fact remains that, during dreaming, the synthesis of new proteins is usually as low as it is during wakefulness: while some limited synthesis may occur in REM, most protein is manufactured in the non-dreaming state.[13] Meanwhile, rehearsal of relevant behaviours would be pointless, or at least not very effective, since subsequent recall of dreams can often be shaky: there's no guarantee you'll remember what you've 'learnt'.

However, another implication of dreams as central to the consolidation of memory and the rationalization of emotions is that this offline option of processing events and experiences that have passed should be particularly valuable for the complex brains of those animals that exhibit particularly high levels of REM sleep.[14] Taking data from eighty-three species, scientists have found that animals with big brains relative to their body size need a conspicuously higher percentage of REM sleep,[15] which suggests that dreaming may indeed contribute to intelligence and cognitive function. Yet others argue that REM sleep measures don't match up with either brain size or complexity: many animals with relatively modest cognitive powers nonetheless display substantial REM, whereas many species with complex brains generate little or none. It seems, therefore, that REM-related dreaming is not some additional and sophisticated add-on to the human cognitive toolkit but, if anything, quite the reverse – a basic default in brains of all shapes and sizes. However, if so, it is hard to imagine what simpler brains, and especially the newly minted brains of human foetuses, would actually dream about, and how different it would be from their waking reality. Suffice it to say that, consistent with this dissociation from mental prowess, dreaming is well known to be dominant in the foetus and infants yet declines in the adult, who surely has developed a greater capacity for cognitive functions and the need to resolve daily problems.

Jerry Siegel, the head of the Centre for Sleep Research, and his group at UCLA have suggested an alternative reason for dreaming that squares this circle: they point out a close correlation between high levels of REM sleep and whether a mammal is 'altricial', as opposed to 'precocial'.[16] Altricial animals, such as cats, rats and humans, are born completely helpless and unable to care for themselves, compared to precocial animals, such as horses and guinea pigs, who are more able to fend for themselves from an early age. While the more readily independent animals have less REM from the outset, and little variation in it over their lifetime, the needy animals tend to have more REM, which then decreases as they mature. So, it seems that immaturity at birth could be the best predictor of REM sleep time in the life of particular species.

THE ONTOGENY AND PHYLOGENY OF DREAMS

By about seven months in the womb, a human foetus is spending most of its time asleep,[17] cycling every twenty to forty minutes into REM, and then even in the newborn this pattern comprises at least half of their total sleep time.[18] This excessive amount of apparent dreaming in the foetus continues on after birth and is roughly double that of adults, in whom a quarter of total sleeping is estimated as REM.[19] So, whatever dreaming and REM signify about brain states, it is something very basic, dominant early in both individual development (ontogeny) and also in evolutionary terms (phylogeny), in both the foetus and in simpler animals. Surely then, the idea[20] that the purpose of dreaming is to consolidate complex cognition is at odds with the disproportionate amount of REM seen at a stage in life, and especially within the restricted environment of the womb, where there will be few life issues to resolve. Aside from the ineffectiveness of REM itself as a candidate biological mechanism for consolidation of memory and cognitive collateral accrued during the day, it would be hard to explain why, in situations in which more modest brains have less need to consolidate a memory, they have the greatest apparent opportunity – namely REM – to do so.[21]

THE NEUROSCIENCE OF DREAMS

In order to get to grips with dreaming, we need to go back to basics: to the brain itself. Let's try to pin down any obvious differences in the underlying brain processes that characterize these two very similar and yet intriguingly different types of consciousness: daytime wakefulness and night-time dreaming. The most fundamental is the relationship, or otherwise, between the inner world of the individual brain and the external environment. How might the brain enable the dreamer to be temporarily disconnected from their immediate physical environment?

Muscles are paralysed during dreams, which explains why a common nightmare involves the frantic desire to run away while being inexplicably fixed to the spot. And, on admittedly extremely rare occasions during wakefulness, when stimulation is unusually and intensely powerful – say, when laughing intensely, or in mid-orgasm – some people can suddenly fall into a complete dreaming sleep![22] However, this mechanism can be occasionally experienced in a more muted form of immobility when we are awake, at times of great excitement or fear, where you may feel unable to move because your leg muscles have relaxed, hence the expression going 'weak at the knees'.[23]

It's not just humans who become paralysed with an extreme emotion such as fear: freezing can most readily be an adaptive response designed to limit movement to protect against predators, such as when a rat smells a cat. Somehow, a powerful and highly charged emotional state, for which nature deems it best that you're immobile, also correlates in extreme cases with a dreaming state and, more commonly, vice versa: dreaming will entail immobility – the inability to escape some approaching danger. One possibility is that in cases of extreme peril or excitement a person will not be able to take suitable action and it is therefore better that they do nothing until they can respond with some appropriate response, or at least evaluate their circumstances. Bear in mind, however, that primitive animals, who are unlikely to have the cognitive capacity to think through such a problem, can behave in this way – so the mechanism was probably there first and was then adapted by evolution to the more sophisticated needs of our species. What could that mechanism be?

One theory is that, during sleep, a frontal brain area deactivates the motor system. As a result of having this route blocked, cerebral activity is somehow diverted backwards to stimulate the posterior brain areas linked to perception, but without the normal sensory input.[24] Then again, there's no reason why this 'activation moving backwards' couldn't also happen in non-dreaming sleep: so, unless there is yet another, and still unidentified, constraining switch specific to dreams, this account doesn't really help with what makes dreaming such a special process.

An alternative scenario is that the primary relay station for the incoming senses – the thalamus – slams down like a neuronal portcullis during sleep, so that sensory signals just cannot make it through. However, we all know that the shrill sound of the alarm clock *does* overcome any would-be barrier. So the temporary yet complete exclusion of the senses from the brain as the defining criterion for a dream doesn't hold. Perhaps, after all, the senses are just an optional extra for an already buoyant consciousness . . .

Rodolfo Llinás, a neurophysiologist at New York University, and his colleague Denis Paré turned the tables and have proposed just that. They have suggested that, for normal brain function, sensory inputs are merely an add-on luxury that happen to be pressed into service in waking hours but are irrelevant to dreaming.[25] Llinás and Paré went on to claim that, in all other regards, wakefulness and dreaming are essentially equivalent brain states, most likely based on a specific neuronal circuit – the much-cited and ever-popular thalamocortical loop. In fact, they came up with a more elaborate version by suggesting that there are actually two distinct such loops: one has a non-specific energizing, or arousal, function, while the other furnishes specific content, which would include that from the senses. So, basic consciousness is never in any case driven from the outside but is a fundamental, intrinsic feature of the brain, merely modulated on occasion by the senses and the external, additional information they happen to convey. Why else, argue Llinás and Paré, might the all-important thalamocortical system have only a minor part of its circling connectivity linked to the transfer of direct sensory input?

Yet we have already seen that there is no reason to suppose that the thalamocortical loops are crucial for consciousness.[26] Although it's an intriguing idea that the dreaming state would be the default setting for consciousness in general, the neuronal infrastructure on which this idea is based just doesn't stand up. First, it is very unlikely that one thalamocortical loop (or even two) is sufficient for consciousness: after all, a dish containing a slice of disembodied thalamocortical circuitry could hardly have a subjective, inner experience. Secondly, arousal on its own doesn't in any case equate with being conscious: rhythms of arousal can be generated in patients who are brain dead.[27]

It would probably be unfair to compare directly the very different theories of Llinás and Paré with those of Hobson and Friston, since they aspire to different goals: the former offer a neurophysiological *description* of the dreaming state, while is the latter suggests more of a functional and abstracted *explanation*. However, neither theory is completely persuasive because neither scenario really explains all aspects of dreaming, such as the ontogenetic-phylogenetic profile of greater dominance favoured in the evolutionary helpless, or the key participation in dreaming of other brain areas, beyond the seemingly merely tangential role of the thalamocortical loops.

For example, another candidate brain site for involvement is the anatomical junction of three areas of the cortex: parietal, occipital and temporal (POT),[28] an area towards the back of the brain linked to the generation of mental imagery.[29] Damage to the primary areas of the cortex – say, those involved in vision – will result in a reduction in the respective visual aspects of dream imagery, but it is only a lesion in this functionally more sophisticated POT area alone that will completely abolish dreaming. However, simply knowing *where* something might be happening does not tell you *how* or *why* that process is occurring. It just does not seem plausible that a simple inhibition or activation between one brain region and others could act as an omnipotent switch cutting off all input from the outside world and turning the brain in on itself. After all, the undeniable puzzle is that the inner world – through reflecting the life and experience of the individual dreamer – provides a very different subjective experience to that of real, waking life.

DO DREAMS REFLECT REAL LIFE OR INNER FANTASIES?

So what is actually happening physically in the dreaming brain? Why and how are dreams subjectively so different to the experience of being awake? Are they an echo of what we've previously lived through in reality, or a new fantasy world conjured up completely internally and, as such, independent of the outside world? In neuro-speak, this dichotomy could be rephrased by asking whether dreams are, effectively, a bottom-up neuronal process linked closely to neuronal mechanisms of primary perception, or a top-down phenomenon more closely reflecting a sophisticated imagination.

Here scientific opinion divides. In one camp, sleep expert Allan Hobson argues that the phenomenon is generated bottom-up, thanks to the fountain of acetylcholine unleashed on higher centres without the restraint of the other modulators in its chemical family: in this case, dreams would be a basic if somewhat distorted and simplified form of perception. And, in many ways, this seems correct: dreams *do* echo perceptual experiences in real life. For example, patients with face-perception problems don't dream of faces.[30] In the same vein, those who lose their sight after the age of seven can still conjure up visual imagery in their dreams,[31] presumably because their earlier visual experience has also influenced their subsequent representations of the environment so they still have the potential ability to produce visual imagery. Meanwhile, the fact that those who have been blind from a very young age do not 'see' in their dreams[32] yet again suggests that the repertoire available to the dreamer corresponds strongly to their waking lifestyle and experiences: a bottom-up process. What, then, of the kind of time lag we have in our brains, where you might dream about something quite a while after it has happened, or continue to dream about people long after an encounter with them? Presumably, these memories, though less recent, would constitute the same mindset as that for events which occurred, perhaps, only yesterday.

In any event, in order to investigate this bottom-up scenario more thoroughly, researchers in Kyoto have attempted to 'decode' the visual content of dreams. Every time the brain wave pattern (EEG) of their subjects showed a particular profile, they were woken up and asked to report the visual experiences they had just been having while still asleep. In this way, over two hundred dream reports were collected, with a total of thirty to forty-five hours of experiment time for each volunteer. Although there were some florid exceptions, like meeting famous celebrities, most of the time the subjects reported dreams relating to their daily life.[33]

But the feasibility or otherwise of the content actually wasn't that important. Rather, it was merely the key concepts that featured: the strategy was to search for terms that appeared frequently in the reports, such as 'female', 'male', 'car' and 'computer'. Then, by showing photographs of the appropriate objects to the subjects while they were awake, the researchers could derive a kind of individual brain-scan signature for each object which they could then compare with the scans just before the participants were aroused from their dream. It turned out that each specific visual experience during dreaming correlated with the brain scan activity patterns shared by everyday stimulus perception: imaging patterns in real, waking life matched up specifically with the literal contents of dreams, and with enough accuracy to be able to predict what the dreamer would report. So, this work supports the idea that dreaming and everyday perception may share 'neural representations' – neuronal networks – in the higher visual areas. It would seem that dream content does indeed reflect individual brain-cell networking from the bottom up, as a result of previous sensory experiences.

But then there's the opposing view: since the time, famously, of

Sigmund Freud, it has been countered that dreaming is actually top-down, emanating from the more sophisticated areas of the brain that are responsible for our most exotic and idiosyncratic inner thought processes. Freud's theory of dreams was based on his tripartite division of the mind into the Id (the source of atavistic urges for destruction and procreation), the Ego (which interprets these urges into specific and plausible contexts) and the Super-ego (which applies an additional and restraining moral filter). Freud suggested that dreaming unmasks the Id, and provides a space where its covert drives may be acted out. However, these desires may be so disturbing that the mind translates the troubling content into a more acceptable symbolic form. As a result, weird and incomprehensible dream images occur. Moreover, according to Freudian thought, the reason memory about dreams is so unreliable is because the Super-ego is at work protecting the conscious mind from the harsh reality of its own subconscious. So the idea here is that dreaming is top-down: generated by sophisticated mental processes coming from within the psyche, not a straightforward means of processing of the humdrum outside world.

In support of the top-down theory of dreaming, there's a clear contrast between levels of complexity in dreaming and waking states in children, compared to adults. Studies have uncovered the fact that pre-schoolers' dreams contain no feelings or social interaction: they are static and plain scenes in which the child plays no part but observes, for example, a horse eating.[34] If young children's dreams were directly perception-driven, then they would be as animated, and at the same cognitive level, as perception itself in their waking life.

Another reason there might not be a direct connection between the consciousness of dreaming and that of wakefulness comes from clinical neuroscience itself, in certain cases in which brain damage seems to have put an end to dreaming altogether. In 100 per cent of all reported instances, the damage was specifically located in the forebrain. So, if the cortex is pivotal, this suggests that dreaming is not linked directly to bottom up, incoming perceptual experiences but is triggered by 'higher' centres such as the frontal cortex, as well as the tripartite cortical junction, the POT, mentioned earlier. The most plausible anatomical location for dreaming is therefore in the *interaction* between the frontal cortex and the POT.[35] However, ever since the mid-twentieth century, yet a further sector of the cortex has been implicated, too.

The Canadian neurosurgeon Wilder Penfield, whom we met back in Chapter 1, pioneered the technique of direct brain stimulation in awake patients suffering from epilepsy.[36] His work is of particular relevance here because he discovered that, when the exposed brain surface of yet another cortical area (the medial temporal lobe) was stimulated, patients occasionally would report that the procedure triggered memories, but ones they said were 'like a dream'. There were no specific time and space coordinates which would normally, as in waking life, enable the individual to place such an episode in the wider context of the day, month or years of their ongoing life story. Rather, they were more seemingly generic occurrences lacking – as in dreams but unlike in real, waking life – abstract thought, calculations and future plans.

The 'dream-like' experience, as we all know well, is one of blended characters, broken and illogical narrative and a disconnect with the environment: small surprise, if so, that a significant difference between dreaming and waking life is that the memory of a dream is usually feeble by comparison with the robust recall of daily existence. And yet just such a mindset characterizes clinical psychoses such as schizophrenia, in which the sufferer (or dreamer) is in an almost delirious state, disorientated and disconnected from the immediacy of the banal, grounded narrative around them. In both dreaming and schizophrenia, logic and reasoning are severely compromised and the individual loses their grasp on reality. In both cases, flimsy attention veers from one person, object or disconnected event to the next, with a perspective devoid of insight and self-consciousness and frequently characterized by delusions and hallucinations.

An evaluation of individual consciousness known as the mental status exam reveals that accounts of dreams, and accounts from schizophrenics are indistinguishable from one another,[37] though a crucial difference is that paranoia is a core symptom of schizophrenia yet largely absent in dreams. Then again, although in the dream world there is no elaborate context, no clear narrative containing well-defined individuals whom you are convinced are plotting against you, there *is* frequently a sinister feeling of ill ease, of danger lurking somewhere or of some shadowy adversary who is out to get you.

Dreams are characterized by strong emotions but emotions of a particularly active type only, such as anger, fear and joy, in contrast to those flatter, passive emotions which are less immediate and more dependent on values and assumptions: sadness, shame and remorse. What is the crucial difference? One possibility is that the 'active' emotions do not depend so greatly on a pre-existing context, an established neuronal network in the brain: a small child, or even an animal, will experience them as either positive or negative reactions to a here-and-now situation. In schizophrenia, the situation is similar, in that an elaborate context is less prominent and less robust, accounting for the lack of logic and inability to interpret proverbs that characterizes schizophrenic thought, along with reactivity to whatever comes along – as with children. In contrast, only older children and adults can experience the more context-dependent 'flatter' emotions, because only their brains are sophisticated enough to provide the requisite neuronal infrastructure – the connectivity – to give the background the pre-existing 'meaning' that lends such emotions their particular significance.

So, the basic difference between dreams and the real world is that when we dream there is a disconnect between feelings and the *context* in which those feelings are deemed appropriate. Laughing at a funeral in the real world would be inappropriate because everyone else finds it so: in contrast, in dreams, the norm – the context – is set by the dreamer alone, and only in retrospect, on waking, would be regarded as mentally deranged. Dreams are not so much, then, an accurate replay of daily life but a powerful form of imagination that is somehow indirectly and loosely linked to the dreamer's waking existence. As such, dreaming must reflect hard-wired connectivity of the personalized brain in some way but, then again, not in the same fashion as when a fully awake individual is having an ordinary ongoing experience in the outside world.

In order to explain this difference, it seems that bottom-up versus top-down perspectives are equally unhelpful, not least because both seem to be right. Instead, we need an account that reconciles the self-evident hard-wired personal networking reflecting everyday waking life with the very different schizophrenic-type content of the subjective dreaming experience.

Schizophrenia itself is associated with unusually high levels of functional dopamine,[38] so it may come as no surprise that damage to the pathways which release dopamine lead to a loss of dreaming, while an increase of dopamine release increases the frequency and vividness of the content of dreams,[39] albeit without any change to REM itself. Conversely, anti-psychotic drugs that are prescribed for schizophrenia and act as dopamine blockers also reduce excessive and vivid dreaming.[40] What might be happening? The forebrain – specifically, our old friend the prefrontal cortex – is the only part of the cortex to receive a rich supply of dopamine from the fountain in the deeper areas. So, if there's an excess of dopamine, which we know inhibits the prefrontal cortex, then the brain will be in a similar basic state comparable to that of schizophrenia, a vivid, emotional world that exhibits reduced logical reasoning.[41] The crucial event is that, this time, dopamine inhibits the prefrontal cortex[42] – a situation which is associated with dreaming.[43] But wait: we saw earlier that damage to the prefrontal cortex leads to the exact opposite – a loss of dreaming.[44] So how come that, now, an under-activity in the very same area enhances it? Clearly, the active inhibition by dopamine of living cells in the prefrontal cortex must be a different situation altogether from those cells being completely dead.[45]

However, there's more to dreams than dopamine. A fascinating clue into a wider neurochemical landscape is that pain is usually suppressed while we're dreaming,[46] which explains the 'dream-like' state described by those taking the painkiller morphine (named after the Greek god of sleep). Morphine is an effective painkiller due to an entirely different transmitter system: the opiates. How do they fit into the picture?

As we've seen repeatedly in different instances throughout a typical day, neuronal assemblies provide a useful Rosetta Stone for relating objective physiology to subjective experience. In the previous chapter, I suggested that the degree of subjective pain felt could be linked to unusually large assemblies, and reduced by the inhibitory action of opiates. It follows, therefore, that if alleviation of pain by morphine were to reduce the size of an assembly, then, correspondingly the characteristic 'dream-like' subjective experience that ensues would also imply that the similar experiences of real dreams, too, could be linked to small assemblies. Interestingly enough, schizophrenia, a condition in which there are clear parallels to dreaming and indeed the small-assembly state, also involves a higher threshold for pain.[47]

DREAMING: A CASE OF SMALLER ASSEMBLIES?

A common link between children[48] and schizophrenics[49] is that their brains are characterized by high levels of dopamine at work, and, in turn, with an under-functioning prefrontal cortex. We have already seen that excessive dopamine and an underactive prefrontal cortex can be linked to small assemblies, so surely the mechanisms at play here, whatever they may be, could just as easily apply to dreams.

What if the basic physiological correlate of the phenomenological experience we call dreaming is one of unusually small assemblies? In order to test this theory, we can see if it's now possible to use the stone-into-the-puddle analogy to account for various facts and reconcile various unexplained phenomena and anomalies that have been cropping up in the different theories of sleep and dreaming.

First, the stone-into-the-puddle assembly model might explain why the sense of smell doesn't feature in dreams.[50] Smell, as we saw earlier, more than any of the other senses, depends not on any residual internal activity, as would be the case, say, for vision or hearing, but instead is completely a function of the force of the stone throw: this stimulus therefore, driven as it is externally, is completely absent in dreaming.

Second is the large amounts of time spent dreaming by the human foetus and in the young of species that do not need to engage immediately with the outside world: for the foetus,[51] locked in the womb and with minimal sensory input, arousal levels will be persistently modest: the viscosity of the puddle will impede the ripples, while the force of the stone throw will be weak and the size of the stone small, given the paucity of neuronal connectivity. Dreaming, namely small assemblies, could well be the default mode. Once born, in species that do not need to process inputs from outside and interact efficiently as much as those who are mature at an early age, again, the stone will not be forcefully thrown: assemblies will be relatively small and the consciousness of full wakefulness will not be so necessary.

Third, there are the deleterious effects of sleep deprivation, including prevention from dreaming: it is not so much that crucial chemicals cannot be stockpiled (which, as we saw, didn't happen in dreams anyway), but rather that the brain is fully conscious and awake – specifically, it is continuously receptive to the outside world. Constant wakefulness entails a continuous sequence of stones being thrown into the puddle, such that the time-consuming plasticity for establishing long-lasting neuronal connectivity, namely larger stones, will be harder to form, due to constant competition. The stones will all therefore end up smaller, increasing the likelihood that the ensuing assemblies will be small – a state which, as we have seen, is characterized by heightened emotion, a lack of logic and even mental illness.

Fourth, recall the studies showing that REM and dreaming itself can occur independently of each other and that, even without brain damage, people can experience dreaming without having displayed the characteristic rapid eye movements. So, in terms of the brain, what could these two types of dreaming (non-REM and REM) have in common that still nonetheless differentiates them as underlying a subjective dreaming experience, as opposed to full-blown wakefulness? The common feature could be quantitative rather than qualitative, namely, that in both cases the assemblies are far smaller than when awake. But because an assembly is continuously variable, rather than all or none, this same sliding scale could then be used to differentiate non-REM from REM dreams further. If the REM state is due to the promiscuous release of the modulating acetylcholine from the brain stem, then it will access all areas of the cortex, giving the opportunity, therefore, for relatively extensive assemblies. In contrast, if this system is not activated, as in non-REM dreams, then the dopamine that we know is still released will be able to act only on its much smaller, selective target: the prefrontal cortex. Irrespective of other factors in determining how and where the crucial assemblies are formed, the absence of an additional modulator, such as acetylcholine, would make it likely that they would be even smaller. Dreaming could still be possible, albeit rarer.

Fifth, the actual length of the stages is more marked by REM changes throughout the night. Neuroscientist Bill Klemm from Texas A&M University has put forward the intriguing proposal 'that the brain uses rapid eye movement to help wake itself up after it has had a sufficient amount of sleep'.[52] His idea is certainly supported by clear facts: as the night goes on, episodes of REM become longer and more frequent as morning approaches. It is as though we are trying to wake ourselves up throughout the night by engaging REM, and are more successful as time goes on, as the REM periods then lengthen (see Chapter 2). Eventually, the REM is significant enough to pull us completely out of sleep.

The relevant anatomy supports this theory, as the cortex feeds back to the primitive brainstem, in which the release of the transmitter acetylcholine is triggered[53] that, in turn, regulates the rapid eye movements. So, if REM sleep specifically increases in duration throughout the night and if, compared with non-REM, it is linked to relatively larger assemblies which become ever more extensive as time, and the night, wears on, the border with an REM/dreaming assembly and an initially small assembly for waking will become increasingly blurred. If assembly size functions along a continuum, then not only would non-REM dreams be merely quantitatively different, in assembly size, from REM dreams, but a similar phasing of REM dreams into wakefulness with a growing assembly would also be plausible: the crucial issue, in terms of the level of consciousness you are experiencing, would be down to final assembly size alone, which would defy a rigid compartmentalization into dreaming versus non-dreaming states. The size of the stone at any one time – namely, existing brain connectivity – will determine the extent to which daily life is abandoned in dreams for less logical and more exotic scenarios or whether the dreaming experience is more reflective of reality.

This scenario could also explain 'lucid' dreams, a state which has been acknowledged since the time of Aristotle.[54] The Greek philosopher claimed that 'often when one is asleep, there is something in consciousness which declares that what then presents itself is but a dream': in essence, the dreamer is aware that they are dreaming. The subjective consciousness occurs, therefore, on a continuum between normal dreaming and waking: arguably, this more sophisticated, self-conscious content would suggest that the lucid-dreaming assembly was larger still and enhanced still further the probability of waking.

In line with this hypothesis is the phenomenon of a 'wake-initiated lucid dream', in which the dreamer transitions from a normal waking state directly into a dream state, with no apparent lapse in consciousness. The sleeper apparently enters REM sleep with unbroken self-awareness directly from the waking state.[55] If REM-based assemblies are larger than non-REM, and lucid-dreaming assemblies larger still, then it will mean that they might be nearer to reaching the size of assemblies that characterize wakefulness; but, more importantly, that assemblies are linked directly to different degrees of consciousness, including different degrees of dreaming: the larger the assembly in a dream, as it approximates more and more to wakefulness, the more likely that it will reflect the content and reality of the world you experience when you are awake. Hence, the assemblies of all types of dreams will be smaller than assemblies in wakefulness, but there will be a continuum spanning both top-down inner fantasy to bottom-up everyday experiences, depending in turn on the size of the stone – the degree of neuronal connectivity activated.

Assemblies are analogue, endlessly dynamic and the end product from moment to moment of a range of different factors. As such, they can link micro-level cells and synapses with macro brain regions and thereby provide a framework for understanding the phenomenon of dreaming, how it varies in content, and its relation, or otherwise, to the waking world. But, by now, you have drifted still further away from that world, into a deeper, dreamless sleep . . . oblivious now, as the night unfolds.

8
Overnight

It's been a long day, but now it's finally over. The succession of different mental states you have been through can now be expressed bilingually – with the terminology of objective physiology and the simultaneous, corresponding language of subjective phenomenology. The degree of neuronal activity, on the one hand, is matched by the impact of the senses on the other. We could talk objectively about your pre-existing associations or, in more subjective terms, personal 'significance'. Let's also factor in the availability of modulators such as dopamine – the physiological parallel to the subjective sensation of arousal. Yet another objective feature is the formation of competing assemblies, along with its first-person corollary, distraction, manifest as levels of assembly turnover: meanwhile, the linking of the passage of time, a narrative of past, present and future, could relate to the degree of activity of, and contribution from, the sophisticated prefrontal cortex.

These different pairings, in each case, effectively two sides of the same coin, mean that we can now switch back and forth between the neuroscientific features of an assembly, while constantly cross-referencing subjective phenomenology. We might start with the physiology and see how a similar outcome – say, an unusually small assembly – can arise from different combinations of various physiological factors, such as a paucity of neuronal connectivity, high levels of the modulator dopamine or a fast-paced stimulation. Then, at the same time, we can travel in the other direction and interpret everyday activities, or familiar types of mental conditions and hence of consciousness, such as childhood or schizophrenia, as corresponding, in turn, to the dynamics of particular profiles of assembly.

Most importantly, throughout the journey through your day, we've seen how the role of neuronal assemblies as an intermediary between objective and subjective states might explain various puzzles: human versus non-human consciousness; differing levels of anaesthesia; the efficacy of an alarm clock; the distinction between hearing and vision; the nature of dreaming and even, as with the effects of environmental enrichment, the evolutionary survival value of consciousness itself.

ASSEMBLIES: A ROSETTA STONE LINKING PHYSIOLOGY AND PHENOMENOLOGY

So now various kinds of prediction are possible (see Table 2), as to how these first- and third-person states might link to each other in terms of the common factor, the Rosetta Stone of net assembly size. On the upper bar are the different factors, expressed in terms of objective physiology, which can all vary independently of each other, such as stimulus, neuronal connectivity or chemical modulators; on the lower bar, the very same factors are expressed in terms of subjective phenomenology.

How might such a table be helpful? Let's take one example: the big difference between depression, say, and anxiety, is that depression features a continuous theme or scenario, for instance, an unchangeable situation such as the death of a spouse, or a general, protracted chemical state: a persistent flattening of general mood. In contrast, anxiety generates multiple, fast, imaginary scenes of terrible outcomes as though they were occurring in the real world: an anxiety over mortgage payments might conjure up a vision of going to court, of losing your home, of losing your spouse, and so on, and so on.

So, while both depression and anxiety would be based on an *internalized*, extensive neuronal circuitry – a large stone – the ensuing arousal levels would be different in each case. Levels of the modulating fountains are lower in the depressed, while in those who suffer from anxiety they are higher, as in fear, but, then again, anxiety would now differ from fear, as fear is dependent on strong *external* stimulation, necessitating, this time, a smaller stone thrown forcefully. With anxiety, therefore, the turnover of neuronal assemblies

Neuronal connectivity (size of stone)	Trigger (force of stone throw)	Levels modulator (puddle viscosity)	Assembly turnover (frequency of stone throws)	Net assembly (size extent of ripples)	Daily activity/ brain state
Sparse	Strong	Low	Low	Small	Alarm bleep
Decreasing	Absent	Low	Absent	Ever decreasing	Anaesthetics
Sparse	Strong	High	High	Small	Childhood/animals
Sparse	Strong	High	High	Small	Fast paced sports
Extensive	Weak	Medium	Low	Large	Jogging
Sparse	Strong	Medium	High	Small	Taste/smell
Extensive	Weak	Medium	Low	Large	Vision
Sparse	Strong	Medium	High	Small	Hearing
Sparse	Strong	Medium	Low	Small	Toch
Sparse	Strong	High	High	Small	Music: rave
Extensive	Weak	Medium	Low	Large	Music: classical
Extensive	Weak	Low	Low	Large	Working in office
Very extensive	Strong	Low	Very low	Large	Depression
Very extensive	Weak	High	High	Large	Anxiety
Sparse	Strong	High	High	Small	Fear
Very extensive	Strong	High	Low	Large	Pain
Extensive	Strong	High	High	Small	Schizophrenia
Sparse	Strong	High	Low	Small	Alzheimer's disease
Sparse	Weak	Low	Low	Small	Alcohol
Extensive	Strong	Medium	Low	Large	Abstract thought
Extensive	Strong	Low	Low	Large	Meditation
Extensive	Weak	Low	Low	Small	Dreaming
Significance	**Degree stimulation**	**Arousal**	**Time perception**	**Degree of consciousness**	**Phenomenology**

would be as high as in fear but primarily driven by internal factors, as in depression: all three distinct states could consequently be distinguished by differential contributions of the various factors that give rise to the final emergent assembly, and hence consciousness.[1]

Heretical though it may seem, the crucial point here is not so much whether the predictions of assembly size listed in the table opposite turn out to be right or wrong, but that such predictions might eventually be *testable*. What neuroscience can bring to the party is not so much the answers but informed questions that can be investigated empirically: in Karl Popper's famous terminology, a 'falsifiable hypothesis'.[2] However, in order to go beyond validating the role of assemblies in dealing with the all-elusive subjective phenomenology deriving from the holistic brain, we would need human subjects.

Back in Chapter 2, we met Professor Brian Pollard, of Manchester University, who has developed an innovative new technique for studying the brain known as Functional Electrical Impedance Tomography by Evoked Response (fEITER). This technique allows his team to look at the brain not only over very short time scales but also in a non-invasive way that opens up the opportunity for testing in humans.[3] While equally non-invasive techniques such as fMRI are comparably painless and practical, they will only ever monitor indirect parameters, such as changes in blood flow, while fEITER reads out a direct measure of changes in brain states, namely, changes in neuronal electrical resistance:[4] this means that, for the first time, there is the potential for monitoring the human brain in real time. You can imagine our delight when Professor Pollard reported in the press that his data so far appeared to support our lab's approach to 'the nature of consciousness itself'.[5] Using fEITER, it might then eventually be possible to test out the kinds of predictions shown in the table. Such studies would give powerful insights into the neuronal mechanisms underlying various subjective states of mind.

But the profile of assemblies that, even then, it might be possible to capture and analyse will only ever be at best an *index* of degree of consciousness, not a demonstration of the phenomenon itself. We must not forget that assemblies, however precise, will only ever provide a correlation of neural events with consciousness, not a causal link. So what happens next, after an assembly has been formed?

THE BRAIN IN A BODY

An essential and basic fact which we have conveniently ignored so far is that the brain exists within a body – it is not floating free in some kind of surreal vat, as philosophers sometimes like to imagine.[6] Instead, the nervous system interacts constantly and intimately with the immune and endocrine systems, otherwise there'd be biological anarchy: moreover, there would be no accounting for the effect of placebo drugs on illness, of depressed mood on health, or of hormones such as oxytocin ensuring that you feel attached and close to someone. Any realistic theory of consciousness based on neuronal correlates must take into account the incessant interplay of the brain and the body.

The next question then, is how a transient and flimsy coalition of tens of millions of individual neurons could, nonetheless, broadcast a cohesive message out to the rest of the organism and its various control systems. In other words, how could a particular assembly report to all the other organs below the eyebrows, and in return be influenced reciprocally by their diverse machinations?

Whatever kind of signal it was that was sent out from the brain to the rest of the body, it would have to convey not just the size of the assembly in question but its degree of intrinsic activity, the duration of its time window and information about the particular brain region where each was generated. Assemblies have the potential for rising to this challenge, since they could offer a read-out made up of various and highly variable factors; since these factors can be differentially manipulated (as in the table on p. 176), there will be a different net read-out reflecting each individual assembly from one moment to the next.

It is just not feasible that every different quantitative and qualitative factor we've been exploring could ever be replicated exactly the same a second time around: this means that, unlike particular anatomical brain regions, and/or their signature electrical activities, every assembly, wherever and whenever it is generated in the brain, will be unique – a one-off that makes it, compared to other possible neural correlates of consciousness (Chapter 1), a much more appropriate candidate to be commensurate with a unique moment of consciousness. But if so, we now need a means of sending this combined package of qualitative and quantitative information in a way that could also have an impact onwards on non-neural systems and peripheral organs of the body – say, the gut – as well as on the other great control systems: the autonomic nervous, endocrine and immune systems. Some kind of general system interfacing must exist whereby the peripheral organs and body processes keep in close communication, back and forth, with the brain.[7]

Luckily, the perfect go-betweens exist: peptide molecules. Peptides are made up of the same building blocks (amino acids) as proteins, but are distinguished from them in terms of size – they can be much, much smaller. The term itself derives from the Greek for 'digest', since these compounds have long been associated with the gut, although we're about to see that they can also operate as powerful transmitters in the brain. In fact, the gut and the brain appear to be in close dialogue, an interaction long acknowledged unwittingly in the phrase 'gut feelings'.[8] Cells in the gut release peptides which, in this case, act as hormones. These peptide hormones in turn not only affect local digestion but also exert an action on peripheral nerves, as well as those in the spinal cord: as such they could have a significant influence on brain processes underpinning memory and emotions, with a wide range of brain areas playing a part in this gut-and-brain dialogue.[9] But it's not only the gut that, remote within the body, can talk to the brain via these obliging and versatile emissaries: for example, a peptide triggering an increase in blood pressure – angiotensin – is produced by the kidneys but can also influence sophisticated brain function, such as learning.

Biologists have known for a long time that crucial and urgent information about the body, such as core temperature, glucose and insulin levels, is rushed back directly to key brain regions, while read-outs of blood pressure and heart rate will be sent on into the brain via various mechanisms in the arteries, the heart and specialized autonomic nerves.[10] One imaginative idea is that the eventual integration of all these various incoming signals could be realized as a collective, single, 'hedonic state'[11] of well-being. While this intriguing concept is still frustratingly vague, it's certainly easy enough to envisage other modulators, for example peptide hormones, translating, in subjective terms, into predisposition or mood. But how would any particular state of consciousness such as this then feed *back* again out from the brain into the rest of the body, to ensure constant conversations between the central nervous system and the vital organs, the endocrine and immune systems?

Perhaps the best-known case of an intimate interplay between these great control systems of the body is a process which operates during stress. This all-too-familiar experience is triggered by a specialized hormone (corticotropin-releasing hormone) and the transmitter noradrenaline which, among many other things, combats the inflammation resulting from tissue damage. At the same time, this system is involved in more protracted psychological states: for instance, its long-term activation can predispose people to depression.[12] A wide range of disorders, from inflammatory diseases such as rheumatoid arthritis to less readily definable mental problems, occur when this tripartite organization (endocrine, immune and central nervous systems) malfunctions.

Although we don't know as yet how it happens, we do know that such three-way communication must occur. Depression, for example, increases the chance of illness through a compromised immune system: in one twenty-year study of over two thousand middle-aged American men, those who showed a tendency towards depression had twice the risk of developing a fatal cancer later on, irrespective of other relevant factors, such as smoking or family history.[13] Similarly, a survey in Norway has shown that those with significant depression had a higher risk of dying of most major causes of death, even after adjusting for age, medical conditions and physical complaints;[14] depression also increased the risk of onset of coronary disease.[15]

Research reveals that the immune system can be conditioned with just the same mechanism in the brain that altered the famous dogs of the Russian physiologist Ivan Pavlov: the animals ended up salivating merely at the sound of a bell which had been previously associated with food. In one experiment on rats, a sweet taste came to elicit the same response as an immune-suppressant drug when the food stimulus was eventually administered just on its own. The rats ended up with an immune system suppressed not by the drug at all but by the mere *association* of the effects of the drug with the sweet taste: this taste in itself had the same effects as the toxic drug, and ended up killing the rats.[16]

It makes perfect sense that a particular learnt association, such as the sweet taste, would trigger chemicals such as peptides which would have an effect on both the brain and the immune system. But the big puzzle still remains: how is such an interaction orchestrated? It would be unlikely that local chemical surges in different circuits haphazardly rise and fall with casual disregard for the status of the organism as a whole: the body would then risk receiving a deluge of mixed messages. The holistic output scenario of the brain of 'being depressed' or even the still vaguer 'hedonic state', in short, a state of mind, must somehow have a physical representation. While we cannot as yet describe such a process in precise, empirically based detail, assemblies fit the bill as just the right level of brain organization to correlate with top-down consciousness, while at the same time constituting the collective emergent property of diverse bottom-up brain mechanisms and processes. But what kind of signal could the generation of an assembly send out of the brain to the body? It certainly couldn't be simply electrical – the universal coinage of neuronal networks – since this time the message would be targeting across very long distances and a diverse range of non-neural biological tissue. The message would have to be chemical. Cue the peptides . . .

As well as performing a whole host of functions outside the brain, peptides can work within the brain as transmitters in their own right: the naturally occurring enkephalin, for example, is an analogue of morphine, which functions within the brain to alleviate pain perception. However, a fascinating general feature of pretty much all peptides in the central nervous system (CNS) is that they are frequently co-located in a single neuron, along with a familiar, conventional transmitter:[17] dopamine cells, for example, might also contain enkephalin.[18] Why should nature want to overload a neuron with two different transmitters, if each does the same basic job?

But just suppose this wasn't the case, and conventional transmitters and peptides were *not* doing the same job. We now know that, compared to the release of transmitters on their own, the peptides are selectively emptied out from different sub-compartments and released from the neuron only under certain conditions: the cell has to be much more active, and for a longer stretch of time. Because the hundred or so peptides in the brain[19] can vary in the amount released (a quantitative factor), as well as in their own diverse chemical identity (a qualitative factor), the brain now has at its disposal a powerful additional tool: a single, *digital* parameter, the all-or-none firing rate of action potentials, can now be cleverly converted into an *analogue* parameter, where the additional peptide transmitter is released only when activity continues at a higher rate and for a sustained period. The quantitative amount of peptide released, combined with its very qualitative identity, will reflect the degree to which a certain neuron in a certain place has been active over a protracted period of time. So perhaps nature has not been so extravagant after all in allocating two types of transmitter molecules to a single neuron: one, the classic transmitter, operates at a local, short-term level while the other, the peptide, functions on a larger scale in space and time.

We've seen that one of the most basic features of all assemblies, even those that are relatively small or weak, is that they are all of long duration: usually, hundreds of times longer than a single action potential, so it follows that generation of assemblies would offer the perfect circumstances for peptides to be released. This release within the brain could communicate to other neuronal groups, and indeed to the rest of the body, that a significant assembly had been formed, rather than just an isolated synapse or two being active. Moreover, this information would not be a simple, digital, on–off switch. Instead, the additional qualitative factor, the chemical identity of the peptide in question, could give a highly individualized, one-off read-out about a particular assembly according to: 1) the levels of the specific different chemicals released; 2) the duration for which the release lasted; and 3) the particular combination of various peptides, which in turn would give a clue as to its spatial size and even its anatomical provenance, since different brain areas will have different peptide signatures. In turn, these assembly features would, unlike bottom-up cellular events or top-down macro-anatomical ones, reflect and read out a diverse range of functional information at any one time: arousal, plasticity (memory), sensory input and internal body state (including hunger, pain, and so on).

To recap one more time: in the generation of an assembly of active brain cells, different factors will differentially determine its final maximum size (the extent of the ripples), which is in turn determined by the degree of sensory stimulation (the force of stone throw), the extent of cognitive associations (the stone), the availability of modulators (the viscosity of the puddle) and the turnover of competing new assemblies (the frequency of subsequent stone throws). All these factors would, as we've seen over and over in many different scenarios throughout the day, define from one moment to the next the extent of a unique assembly.

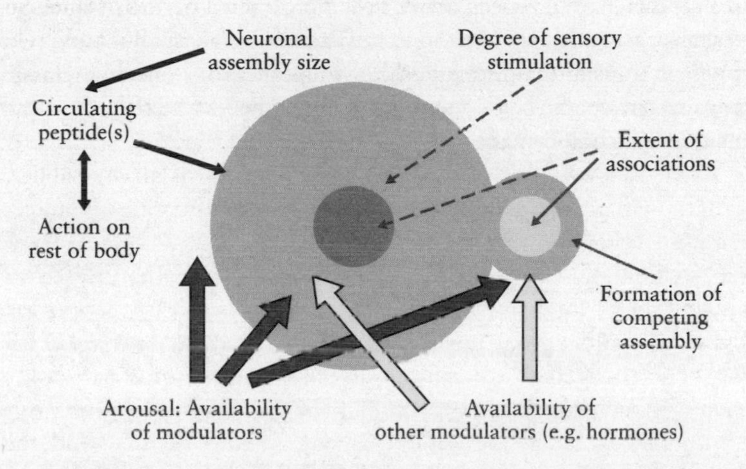

Fig. 9. A possible mechanism for the generation of consciousness. The two sets of concentric circles represent two transient assemblies of tens of millions of brain cells: the largest assembly will dominate at any one moment in the brain, and determine that moment of consciousness. The degree to which cells are recruited, and hence the degree of consciousness, will be determined by a variety of factors, such as the vigour of sensory inputs, pre-existing connections (associations) and degree of competition (distractions), shown by the smaller assemblies starting to form. Signature chemicals, such as peptides, will be released from the transient assembly. The type, number and concentration of these chemicals will thus constitute a unique, one-off signature of the salient assembly in the brain and convey that information to the rest of the body, via the circulation. In turn, chemicals released from the immune system and the vital organs will modify the working assembly of neurons, as will other chemicals such as hormones and the amines that are released in relation to arousal. Consciousness is therefore dependent on the whole body working cohesively.[20]

The simplistic scheme of the way in which the brain and body could subsequently interface via assemblies, shown in Figure 9, will inevitably generate many more questions.

However, for now, it's not that improbable to imagine peptides circulating and acting on the rest of the body: the resultant iteration back and forth between the read-outs from the brain of a unique signature of peptides indicating the existence and status of a unique assembly, in combination with the retaliatory peptides originating from other body systems under specific conditions, will, somehow, translate into moment after moment of consciousness. But how? The problem of what that inner world of consciousness is, and what happens in the brain, hasn't gone away . . . When you wake up, what might tomorrow bring?

9
Tomorrow

When you wake up, you'll have no idea of the length of time you've been unconscious and you may well feel that you have only just dozed off for a few moments, or been comatose for many hours. This familiar experience of disorientation raises a fascinating question: could the fact that sleep involves both a lack of sensory stimulation and a poor perception of passing time mean that the two are in some way linked?[1]

THE PASSAGE OF TIME

Time is the most fundamental feature in our lives. Yet, the more you think about the nature of time and its passing, the more your mind turns cartwheels. Even contemplating the basic unit – a simple, single second – raises difficulties: a second can be defined objectively but in baffling, technical terminology as the duration of 9,192,631,770 periods of the radiation corresponding to the transition between the two hyperfine levels of the ground state of the caesium-133 atom, but even this is an arbitrary gradation.[2]

On the other hand, the experience of time passing is undisputable and a matter of universal regret to all of us: the unidirectional nature linking past to present to future, going forward as an arrow,[3] is the same for everyone. The social anthropologist Alfred Gell has summed up very eloquently that 'there is no fairyland where people experience time in a different way to ourselves, where there is no past, present or future.'[4] And although Isaac Newton believed that time existed in its own right[5] and as a completely independent phenomenon, physicists now view time as linked intimately to space.

But time is not simply something 'out there', as space might more readily be perceived to be. When you wake up from sleep or anaesthesia, you have no sense of the time that has passed while you were unconscious. And, even when you're fully awake, the passing of time can vary enormously, from 'flying' to 'dragging', depending on what you are doing and whether you are enjoying it or not. Time is subjective and therefore inextricably part of consciousness. Just as we create our own inner world, so we create our own time.

Needless to say, the quintessential subjectivity of time lends itself to further unpacking. On the one hand, there is the familiar memory process spanning seconds to months,[6] in which the experience of time is therefore not monitored in the present but is, rather, retrospective. Within this compartment of retrospective time, memory for the duration of an event is a process that is distinct from its actual chronological order. Then on the other hand, there is the direct sensation of time passing,[7] an immediate awareness, on the sub-second to second time scale, described in the early twentieth century by the philosopher-psychologist E. Robert Kelly as the 'specious present',[8] an illusion described as the protracted 'now':

> Let it be named the specious present, and let the past, that is given as being the past, be known as the obvious past. All the notes of a bar of a song seem to the listener to be contained in the present. All the changes of place of a meteor seem to the beholder to be contained in the present. At the instant of the termination of such series, no part of the time measured by them seems to be a past. Time then, considered relative to human apprehension, consists of four parts, viz., the obvious past, the specious present, the real present, and the future. Omitting the specious present, it consists of three . . . nonentities – the past, which does not exist, the future, which does not exist, and their conterminous, the present; the faculty from which it proceeds lies to us in the fiction of the specious present.

TIME PERCEPTION

How might a neuroscientist ever start to gain a foothold on this slippery state of affairs? How might our perception of time be realized in the brain? Inevitably, the easiest strategy is to ask if there is anywhere special within our neuronal networking where time perception could occur . . . But, given the ever-present and constant health warning that we should never assign a sophisticated function to a single brain area, it will come as no surprise that a whole host of regions have been linked to different types of cognitive processes related to time perception.

For example, one strong contender is the cauliflower-like structure at the back of the brain, the cerebellum: it has been called, after all, the 'autopilot' of the brain, coordinating sensory inputs and movement outputs in time-locked sequence. Then there is the 'basal ganglia', a key cerebral circuit for internally driven movements which, again, would be highly time-sensitive. Another possibility is the parietal cortex, a sophisticated area of the brain in which senses and movement are in lock-step. Neurologists have known for a long time, by using imaging, EEG and neuropsychological studies, that damage to the parietal cortex causes people to have problems with spatial tasks and time discrimination,[9] and this area is important for the timing of both auditory and visual stimuli:[10] it therefore plays some kind of more executive role, as does the frontal cortex, which is also implicated in more sophisticated time perception.[11] Once again, such a finding comes as no surprise: damage to the frontal cortex can result in 'source amnesia' where this is not a loss of memory *per se*, but rather a problem with dividing past experience up into clear episodes, parcels of time and space.

Yet any linking of function to structure will depend on exactly *what* aspect of *what* function we are talking about: if the cortex is removed wholesale, then rats, with remarkable resilience, not only remain alive and operational but can still successfully estimate forty-second intervals.[12] So, clearly, some aspects of time perception are more sophisticated, in brain terms, than others.[13] However, lists of different bits of cerebral anatomy will not get us very far in understanding how time perception is possible: instead, let's look not at brain regions as though they were independent mini-brains, but rather at their collective interplay.

Like most complex mental processes, such as vision, time perception is not a simple operation. Just as vision can be divvied up across thirty or so different brain regions – differentially preoccupied with colour, form and motion processing – so time perception can be unpacked by the neuroscientist into different factors that contribute to the final holistic experience. For example, the duration as well as the simultaneity and order of incoming stimuli all affect how we perceive time. So, just as the final experience of vision is composed of different elements that would normally work seamlessly in concert, so too is the experience of time passing[14] Yet this begs the question of just what we mean by 'work' . . .

Everyone studying the neuroscience of time perception agrees that, in the brain, there is no central clock. The American neuroscientist David Eagleman, a leading figure in the field, argues that time is not a unitary phenomenon by pointing to three examples which should dispel the notion that 'time is one thing'.[15] The first is the fact that time can be subjectively different in otherwise objectively identical intervals:[16] surely we all know this only too well when spending an hour bored in a waiting room compared to an hour hanging out in a favoured bistro.

Secondly, if we hear a different, 'oddball' sound (that is, one that is different from all the rest) in a series of repeating and otherwise identical stimuli, we perceive the time duration to be longer between the oddball and the next and/or preceding stimulus. However, at the same time, this expansion of perceived time – say, the pitch of an auditory tone or the rate of a flickering stimulus – does not change concomitantly. This observation again demonstrates that time perception is not a unified, blanket process but is instead composed of independent neural operations that would usually be subconsciously coordinated but can actually be overtly differentiated, as here, under laboratory conditions.[17]

The third example comes from a fascinating and unusual experiment in which time was deliberately contrived to 'slow down', by volunteers being subjected to seemingly life-threatening situations. Most of us, at some point, experience the sensation of time seeming to stutter into freeze-frame – to dilate – for example, during a moment of extreme physical danger or when we have received particularly bad news. If perception of time were, after all, 'one thing', then these protracted moments would lead to a more acute experience of time all round, as though you were watching a video in slow motion, where you'd be able to notice much more detail. But this particular study revealed that this was not the case . . .

Participants were required to fall backwards from a fifty-metre tower into a safety net below.[18] These foolhardy volunteers retrospectively reported, as might be predicted, that indeed the duration of their fall seemed to last longer; as it turned out, 36 per cent reported that their fall was longer compared to the observed falls of the other subjects. But the important point is that there was no evidence that these subjects had an expanded, more detailed experience of time *during* the fall itself: they were unable to report any additional aspects of the world around them. In contrast, the specious present, in which you are experiencing time passing in the here and now, would have more detail – as in the current fashion for mindfulness.[19]

Eagleman concluded from the results observed with the falling volunteers that, since memory is being used for *retrospective* time perception, then more memories or associations can be packed into a frightening episode and this will in turn dilate subjective time perception. In other words, the more memories or associations you have of an event, the longer duration you think it had. Since time perception of the same situation will be different according to whether it's ongoing (the specious present) or recalled, it can't therefore be 'one thing'.

Another possibility is that the passage of time could be, in Eagleman's words, 'encoded in the evolving patterns of activity in neural networks'.[20] The suggestion is that the growth of any neuronal network over time – say, through strengthening of the synapses – will effectively 'code' for that time period itself. But such a scenario just shifts the emphasis from objective events in the outside world to objective events that occur within the brain: there is still the problem of how the *subjectivity* of time perception would come in. The problem remains unanswered.[21]

As I see it, this type of approach based on the concept of encoding, at whatever level of brain function, will always present the same conundrum. If something is encoded then, somewhere or somehow along the line, it will have to be decoded back again to have any value or meaning: that's the whole point of codes. And if consciousness is an inextricable part of time perception, then any notion of decoding lands us slap bang back up against the Read-out Fallacy (see also

Chapters 1 and 4) of who or what is on the receiving end. Instead, a more fruitful way forward may be to lower our ambitions, as we've been doing so far, and simply search for correlates in the brain of time perception, to see if we can then discern some kinds of underlying principles.

Time dilation – an overestimation of how much time has passed – occurs under a variety of different conditions in our everyday lives, but these diverse conditions may share a common underlying neuronal mechanism in the brain. First, greater attention to a stimulus will make time seem as if it lasts longer, for example in mindfulness or, as in the 'oddball' experiments, where processing a deviant stimulus seemed to take longer,[22] or in the contrived 'dangerous' situations.[23]

Secondly, our perception of time passing is actually proportional to the magnitude of the strength of the stimulus: larger/brighter stimuli in higher numbers lengthen the subjective duration of time.[24] A third factor is strong emotion:[25] for example, time is perceived as being longer when angry faces are presented or, more likely in real life, when you're immersed in some intense relationship or activity. The common thread here is an underlying heightened arousal. Fear would certainly be present in accidents and, as we've seen, life-threatening scenarios in general, where time 'stands still'.

A fourth and very different example is childhood – a period which is widely acknowledged to be characterized by a feeling of having all the time in the world. Children live 'in time', namely, very much in the present. For example, it has been reported that an eleven-year-old estimates the passing of one day as being one four-thousandth of life, but a fifty-five-year-old will give a value of one twenty-thousandth, such that a random day is much longer for a child than an adult. For children with attentional disorders, time passes particularly slowly[26] in the same way that it does for schizophrenics,[27] for whom time is no longer a smooth flow of events,[28] but composed of fragmented incidents in which the senses are heightened and where each demands fuller attention. Stimulants that raise arousal levels also lead to an overestimate of time,[29] while depressant drugs that reduce excitement have the opposite effect.[30]

These various findings suggest that there is a clear overlap of interrelated factors linked to an overestimate of time passing: the significance of arousal, emotion, stimulants and ADHD, all of which implicate, among other things, an overactive dopamine system.[31] In turn, excessive dopamine would lead to suppression of the prefrontal cortex,[32] which we've seen is akin to child-like states, and where there is more likelihood of a more vigorous interaction with the outside world, in turn leading to more immediate information processing[33] – just as in accidents.

Our subjective sense of time is also affected by motion[34] and the complexity[35] of events: so, one idea is that the brain estimates the passage of time on the basis of the pure number of events – effectively the number of inputs that zing into the brain. If so, then that would account for the exact opposite scenario, when there is no input at all: in sleep or anaesthesia, where time seems to have passed in an instant. It is not so much that, when time slows down, you have more opportunity to appreciate all that is around you in the specious present, but rather the other way around: when you are processing more than usual incoming information, time *as a consequence* then seems to slow down, but only in retrospect, when you recall an event. The amount of incoming information that has accumulated drives your time perception, not vice versa.

Each incoming stimulus will have its own unique coordinates in space and time, and the two are impossible to separate. Interestingly enough, children often conflate time and space[36] and are unable, for example, to understand the story of the tortoise and the hare – they cannot grasp the notion that faster is not the same as further. Of pivotal significance in the physical world is the concept of the 'space-time manifold', where the three dimensions of space are combined with the fourth dimension of time:[37] perhaps, then, when the brain processes the all-important incoming information that will set the pace for the subsequent perception of time passing, the spatio-temporal relations of the stimulus in question are still preserved. So, time perception, as far as the brain is concerned, may not be an explicit parameter, not some kind of simple neuro-second, but a pattern of change based on space and time acting together in concert.

This idea is far from new. Back in 1915, Émile Durkheim identified time and space as 'the solid frames that enclose all thought', while in 1953 the neurologist Macdonald Critchley,[38] pointed out that 'pure temporal disorientation . . . occurring independently of spatial disorders, is a rarer phenomenon, for more often, the two are combined'. And in 1975, the psychologist James Gibson stated that 'there is no such thing as the perception of time, but only of perceptions and locomotions'. This has even led the neuroscientist Vincent Walsh to go so far as to propose that, in the brain, 'time, space and number are computed according to a common metric'.[39]

This possibility is directly illustrated in a fascinating investigation,[40] in which subjects made judgements of time passing while observing different kinds of spatial environments. In each experiment, there were three different kinds of model – scale model railways, models of a sitting room and abstract models – each presented on a small or large scale. The scale that was smaller than the real-world object was then linked by the subjects to a compression of their subjective time. Once again, it appeared that the effect on time perception is related to differences in the density of the information to be processed, but this time in environments of different spatial scale. In a comparable experiment[41] subjects observing differently scaled environments underwent systematic shifts in the experience of time. Amazingly, the larger the scale spatially, the longer the time apparently took to pass.

If space and time are interlinked, there will be some intriguing implications for how we see the world. It could be that large-scale spaces, such as lakes and mountains, may lead to a perception of much more time passing than is in fact the case: for example, immersion in natural or grand environments such as cathedrals is a likely way to ensure that time slows down, since such places provoke a sense of awe and calm, which are in turn linked once again to dilation in time perception.[42] So, in order to understand how we perceive time, we need some kind of a space-time mechanism within the brain . . .

ASSEMBLIES IN SPACE AND TIME

Once again, assemblies might fit the bill: we can press them into service as the Rosetta Stone linking physiology with phenomenology and see how well time perception might fit with the stone-into-the-puddle analogy. We've just seen that perception of time passing is driven by the degree of incoming stimuli, and never the other way around. Assemblies would help cater for this unidirectional phenomenon, since they are *never* pre-existing: their one-off dynamics are determined, as we've been seeing, by a constellation of diverse factors, each varying differentially from one moment to the next.

When time passes slowly, imagine a high turnover of assemblies which therefore remain small, as the sensory inputs teem in one after the other: these would be found in the cases of accidents, children, schizophrenics, stimulant users, and so on. We might also add to this list a seemingly paradoxical condition, the state of boredom, in which time passes slowly because no stimulus is significant or powerful enough to generate larger assemblies. The opposite, of course, when time flies, is when we are engaged in a single activity that demands all our focus, like reading a good novel: the assembly turnover is slow. However, the most extreme example, the one that prompted this excursion into time perception in the first place, is that when we wake from sleep or anaesthesia, during which there has been no sensory input at all, there is a complete absence of time passing: it is suddenly morning.

We saw with the experiment of the volunteers falling,[43] that sensory input into the brain cannot be enhanced by the dilated perception of time: so the subjective, *retrospective* perception of time could not be a simple net product of specific assemblies, since they have already been generated and cannot therefore be modified further. If we think about the stone-into-the-puddle analogy, a further factor now comes into play. As well as stimulus intensity (the force of the throw), meaning (the size of the stone), arousal (the viscosity of the puddle), we can add assembly turnover within a specific time frame. This *turnover* of the contributing assemblies will be key in determining the features of an admittedly hypothetical, overarching scenario which links the resulting experience with any particular moment of consciousness.

Activity for any single assembly decays in both time and space after several hundred milliseconds, exactly the crucial threshold found by Libet and others for consciousness to occur.[44] So no single assembly could be a direct correlate of ongoing consciousness itself. But what if the apparent *decay* in each case of a particular assembly was subsequently the all-important factor that triggers the start of something more significant and much more extensive still? This 'something else' would be a kind of holistic single entity encompassing space and time, accounting for the experience of a specious present – namely, consciousness itself. So let's explore how such a global, hypothetical entity – this 'something else' – might be realized. Beyond the highly artificial constraints of the laboratory set-up of anaesthetized rats or brain slices, a much more likely scenario in real life is that multiple assemblies are generated all over the brain then, somehow, they synergize and summate over a certain period of time.

Such a holistic effect could never be detected in the entire 3-D living brain using the current tools we have – not least because we have not formulated as yet exactly what to look for. But, perhaps if we were not encumbered by the trip-wires of practical experimentation, a more unfettered, theoretical science could at least help us to envisage a much more accurate, albeit novel, picture of what could happen in the holistic brain to correlate with moments of consciousness. Let's take a brief look and see how we might model assemblies hypothetically. It would then be possible to go further, to try to work out how multiple assemblies might interact with each other within a given time frame. But, in order to get going, we need to identify the neuronal nuts and bolts that will make even a single assembly possible.

INSIDE ASSEMBLIES

Traditionally, the building block of brain operations is the individual synapse. So, a reasonable starting assumption is that assembly dynamics will be determined ultimately by the familiar process of synaptic transmission. But if an assembly were just to be the net conglomeration of a load of synapses, then we'd run into problems explaining what we actually see, in terms both of time and space . . .

In terms of space, the laws governing classic synaptic transmission would predict a far smaller assembly than turns out to be the case: the assembly boundaries (see Figure 2), extend a distance some three to ten times greater than would have been typically predicted.[45] In terms of time, we know it takes some 300 milliseconds for decay of assembly activity to fall to even 20 per cent of its maximum strength,[46] and that it would take longer still to vanish completely.[47] By contrast, the timescale for traditional synaptic activity would be a maximum of only twenty msec.[48]

A clear contrast between synaptic transmission alone and the additional mechanisms that must underlie an assembly can be seen in the time frames of Figure 10. Note that, while the transmission of a signal from thalamus to cortex, via classic synaptic transmission, takes only five milliseconds to travel some one or two millimetres, a further twenty milliseconds is needed for the assembly to spread within the cortex to its full extent.[49] Something additional is probably happening.

Two important, less familiar types of communication between brain cells could come into play: one occurs on a much wider scale than the synapse (volume transmission), while the other is much more minimalist (gap junctions). Let's take a brief look at each in turn.

Volume transmission is so called because it enables interaction between neurons in a way that is much less specific and significantly slower, yet with the pay-off that it involves far more cells at any one time. This erstwhile revolutionary proposition has been so thoroughly investigated over the last thirty years that it is now accepted

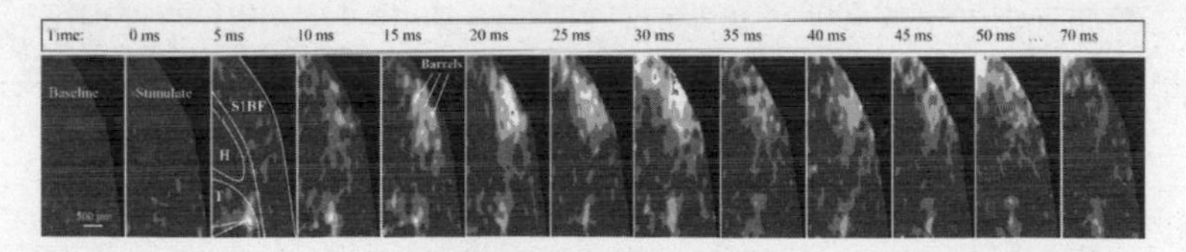

Fig. 10: Neuronal assemblies in a rat-brain slice visualized with voltage-sensitive dyes (Fermani, Badin & Greenfield, submitted for publication). While normal synaptic signalling takes 5msec to travel up to 2 millimetres from thalamus to cortex (see frame showing activity after 5ms), the ensuing assembly takes four times longer to spread some 0.5mm in radius (see frame at 20ms). (To see this scan in colour, see Plate 7.)

as an alternative form of communication and often treated as an equal counterpart to classic 'wired' synaptic transmission.[50]

In fact, it has been known since the 1970s that classic transmitters such as dopamine can be released, as can other bioactive molecules, from parts of the neuron resembling branches of a tree (dendrites) which, typically, have a very different role. Normally, dendrites are thought of as being the receiving zone of the neuron that would act as the target for incoming contacts (synaptic axonal terminals) from other neurons. However, it now seems a general principle that dendrites can release substances in their own right, and do so independently of the electrical blips generated at the cell body. Moreover, release of substances from dendrites is over a much wider area, and is far less precise, suggesting a fundamentally contrasting modulatory process that has very different features to the story of classic synaptic transmission (an action potential firing, impulse propagation along an axon and highly specific release from the axon terminal of a transmitter across a synapse).[51]

But we still have a problem: while synaptic transmission is too local and too fast, volume transmission would prove too slow for the growth of assemblies.[52] There is a strong possibility that a third type of neuronal communication will be at work as well, which might be the ideal complement to the short-lasting, localized synaptic transmission on the one hand and the more prolonged but slow volume transmission on the other.

This third process does not involve neurochemicals of any kind but relies instead on current spreading through 'gap junctions': direct contacts between cells.[53] Interestingly enough, in neuronal networks, very fast oscillations (200Hz) in activity are mediated not via synapses but by current spread through these gap junctions:[54] so, if it's fast oscillations that underlie the sustained activity seen here in assemblies,[55] then, although they'll take longer to get going to reach a maximum, once underway,[56] the spread of activity will reach further than any signal from a synapse ever could. And this extensive spread of activity would be on the same scale that characterizes assemblies in all the various set-ups. Hence, assemblies will be a very appropriate neuronal correlate for the space-time requirements for consciousness, since, unlike localized neuronal circuitry, they are neither hard-wired over time, nor spatially restricted.

These three separate processes at work determining the formation, duration and decay of an assembly in all likelihood will not operate independently[57] and instead will all come into play together. But then the biggest question of all remains: what happens *after* an assembly has been generated to trigger a moment of consciousness? A clue might lie in the timing . . .

A META-ASSEMBLY?

As we saw just now, it wouldn't make sense if a single assembly all on its own was the basis for consciousness: just at the crucial break-point of some 300msec, its signal would be at severely sub-optimal strength of 20 per cent.[58] But, then again, perhaps that's the whole point: what if the 20 per cent decay time at 300msec observed locally in one assembly were itself indicative of the trigger for a more global, long-distance initiation of a more holistic neural correlate of consciousness . . . Imagine a scenario whereby individual assemblies all over the brain could operate independently *up to* about 300msec but, by that time, just as they themselves are decaying, their activity, or rather their energy, has been transferred to some kind of collective pool.

This collective pool – let's call it a meta-assembly – could correspond to one-off holistic brain states, specifically, a moment of consciousness, for the following reasons. Firstly, neural activity appears to contribute to a state of consciousness only when it is sustained:[59] this time window is commensurate with the decay of an assembly, when the electrical potentials recorded in the brain are the same for reportedly 'seen' versus 'unseen' stimuli up until the crucial threshold of 270msecs.[60] Secondly, anaesthetics, which, by definition, abolish consciousness, significantly lengthen the duration of an individual assembly.[61] Thirdly, a time window of approximately this length demarcates the earliest spatial differentiation of different patterns in assemblies for subjective differentiation of sensory modalities.[62] Fourthly, the energy will need to be conserved in some chemical, electrical or thermal form. In the case of heat, pressure will increase, and vice versa: perhaps this could explain why increased pressure, and hence an increase in thermal energy, will lead both to the onset of consciousness in anaesthetized animals[63] and a massive increase in assembly size.[64]

In whatever form it takes, this large-scale transfer of energy will then impact even more extensively on the background activity that characterizes the living brain.[65] This heaving, undulating activity would be readily sensitive to a temporary distortion from the one-off energy transfer, causing global/holistic 'ripples' that could be the real and final correlate of a moment of consciousness. So, any actual neuronal assembly, along with any others generated elsewhere in the brain – importantly, over the same time frame – could now itself become the effective stone in a much, much bigger puddle: a one-off meta-assembly.

But then, where and how could this meta-assembly ever be detected, let alone measured? Remember: there can be no single anatomical location. Rather, we need to envisage a form of integration of large-scale, spatially defined neuronal coalitions that could occur over a time window of several hundred milliseconds – and that means within a kind of neuronal 'space-time manifold'. Since a manifold is a mathematical concept that combines space and time in a single continuum, with time as the fourth dimension, any neuronal meta-assembly is more likely to be described eventually by theoretical physicists than by neuroscientists. After all, by integrating space and time into a single manifold, it is the physicists who have already been able to develop general basic principles and describe in a uniform way the diverse workings of the world from super-galactic to subatomic levels.[66]

However, if a moment of consciousness is indeed ultimately correlated at any one time with a meta-assembly, in turn expressed as a brain space-time manifold, then we would be liberated from the counterintuitive and intellectually unsatisfying task of forever trying to pin consciousness down to anatomically defined bits of brain, which, as we saw back in Chapter 1, took up so much effort in return for relatively little insight. Moreover if time, just as much as space, is now a crucial factor, such a scenario could accommodate what we have been learning in previous chapters about differentiating hearing and vision and differential discrepancies in space-time as well as the subjective passage of time itself, which is perhaps linked to assembly turnover.

The differential chemical, electrical and thermal energy in each case will then, somehow, translate into a moment of consciousness. This 'somehow' is of course the most challenging notion of all, and finally halts us in our tracks. Even if we could model with rigorous and accurate maths not just the assemblies we can see but an invisible, theoretical meta-assembly which might indeed be a final and faithful correlate of consciousness – even then, we would be stumped as to how to proceed to work out a causal relationship and then trace the objective physical phenomena seamlessly to the tantalizingly subjective and personal: a 'hard problem', indeed.

But is this really the place we have to stop? Until we can formulate or articulate what kind of solution would satisfy us, then surely it will be almost impossible to deliver even at the most abstracted mathematical level any kind of answer that goes beyond correlation to causality. Some, like my now sadly deceased father, to whom this book is dedicated, may remain bemused that 'you can't use a knife of butter to cut butter': in other words, that it will always be impossible for the brain to deconstruct itself. Why not give up and work on something of more practical and immediate application, which would have the additional merit of more readily attracting grant funding?

But surely we should not throw in the towel before we even start to explore the issue as fully as we can; surely we should push ourselves to our intellectual limits. And, along the way, even if we have to leave the conundrum of water into wine aside for the moment, neuroscience might make a valuable contribution, one that could and should complement philosophy, psychiatry, psychology, physics and maths. If we combine our efforts towards a multidisciplinary perspective, and stretch the imagination to its limits, then perhaps an understanding of the miracle of first-hand experience will at last become a little less impossible . . . After all, tomorrow *is* another day.

Notes

1. IN THE DARK

1. Blakemore, C. & Greenfield, S. A. *Mindwaves: Thoughts on Intelligence, Identity and Consciousness* (1987).
2. Smith, J. D., Shields, W. E. & Washburn, D. A. 'The comparative psychology of uncertainty monitoring and metacognition'. *Behavioral and Brain Sciences*, 26, 317–39; discussion 340–73 (2003); see also Haynes, J.-D. & Rees, G. 'Decoding mental states from brain activity in humans'. *Nature Reviews Neuroscience*, 7, 523–34 (2006).
3. Crick, F. & Koch, G. 'A framework for consciousness'. *Nature Neuroscience*, 6, 119–26 (2003).
4. Gazzaniga, M. S. 'Forty-five years of split-brain research and still going strong'. *Nature Reviews Neuroscience*, 6, 653–9 (2005).
5. Weiskrantz, L. 'Blindsight revisited'. *Current Opinion in Neurobiology*, 6, 215–20 (1996).
6. Chun, M. M. & Wolfe, J. M. 'Visual attention' in *Blackwell Handbook of Perception* (2000); see also O'Regan, J. K. & Noë, A. 'A sensorimotor account of vision and visual consciousness'. *Behavioral and Brain Sciences*, 24, 939–73; discussion 973–1031 (2001).
7. An example of attention without consciousness. In studies where researchers can contrive a kind of experimental blindsight, the experimenters make use of a highly contrived lab-based technique which psychologists call masking. The experimenter presents to a subject one visual stimulus, the 'mask', immediately after (backward masking) or before (forward masking) another brief visual target stimulus lasting less than fifty thousandths of a second. The result is that, in either paradigm, the subject is unaware of the target stimulus, yet, as in clinical blindsight, they can nevertheless still incorporate the stimulus into their subconscious thinking. It turns out that a subject can be primed for words rendered at the time 'invisible' due to backward 'masking', but only when paying atten-

tion. Similarly, in another experiment, pictures of male or female nudes attracted attention when the subject was not even conscious of the image: the images were 'invisible' to the subject due to a phenomenon called suppression-by-masking. Naccache, L., Blandin, E. & Dehaene, S. 'Unconscious masked priming depends on temporal attention'. *Psychological Science*, 13, 416–24 (2002). Jiang, Y. et al. 'A gender- and sexual-orientation-dependent spatial attentional effect of invisible images'. *Proceedings of the National Academy of Sciences of the United States of America*, 103, 17048–52 (2006).

8. Mack, A. & Rock, I. 'Inattentional blindness'. *MIT Press Cambridge*, 12, 180–4 (1998).
9. Li, F. F. et al. 'Rapid natural scene categorization in the near absence of attention'. *Proceedings of the National Academy of Sciences of the United States of America*, 99, 9596–601 (2002).
10. Reddy, L., Reddy, L. & Koch, C. 'Face identification in the near-absence of focal attention'. *Vision Research*, 46, 2336–43 (2006).
11. Rees, G. & Frith, C. D. 'Methodologies for identifying the neural correlates of consciousness' in *The Blackwell Companion to Consciousness*, 551–66 (2007). A similar effect occurs when a person is first conscious of something and subsequently conscious of something else. The simplest way of contriving this experience is to cause 'binocular rivalry', where one image is presented to one eye and a very different image to the other: the two images compete so that the first one is dominant, then the other, for several seconds at a time. In order to understand exactly what is going on in the brain, similar experiments can be performed with monkeys while the activity of single brain cells is recorded. In one study, the animals were presented with different images of gratings with various orientations so that consciousness 'alternated' between horizontal and vertical grating. Results recording brain activity showed that neurons in a particular part of the outer layer of the brain (the cortex), specifically an area on the side of the head (the infero-temporal lobe), responded only to the dominant stimulus: since this brain region is a sophisticated, less workaday, later-stage processing part of the brain, freed up from the immediate and initial processing of visual inputs, the conclusion ran that consciousness must therefore involve neurons in this part. Meanwhile, the more basic and primary areas of visual processing could be linked to attention, but not to consciousness. This idea would be borne out by the observation that patients who are in a persistent vegetative state and therefore not conscious can still generate evoked activity in the primary areas. It is in the later, or 'higher', stages of pro-

cessing the senses within the brain that we should find clues to the physical basis of consciousness. Blake, R. & Logothetis, N. K. 'Visual competition'. *Nature Reviews Neuroscience*, 3, 13–21 (2002). Sheinberg, D. L. & Logothetis, N. K. 'The role of temporal cortical areas in perceptual organization'. *Proceedings of the National Academy of Sciences of the United States of America*, 94, 3408–13 (1997). Lee, S-H., Blake, R. & Heeger, D. J. 'Hierarchy of cortical responses underlying binocular rivalry'. *Nature Neuroscience*, 10, 1048–54 (2007). Laureys, S. et al. 'Cortical processing of noxious somatosensory stimuli in the persistent vegetative state'. *Neuroimage* 17, 732–41 (2002).

12. It appeared originally in a British comic, *The Beezer*, then later in *The Beano* and *The Dandy*, all published by D. C. Thomson & Co.
13. Felleman, D. J. & Van Essen, D. C. 'Distributed hierarchical processing in the primate cerebral cortex. *Cerebral Cortex*, 1, 1–47 (1991).
14. Velly, L. J. et al. 'Differential dynamic of action on cortical and subcortical structures of anesthetic agents during induction of anesthesia'. *Anesthesiology*, 107, 202–12 (2007).
15. Penfield, W. & Jasper, H. *Epilepsy and the Functional Anatomy of the Human Brain*. (Little, Brown & Co., 1954).
16. Merker, B. 'Consciousness without a cerebral cortex: a challenge for neuroscience and medicine'. *Behavioral and Brain Sciences*, 30, 63–81; discussion 81–134 (2007); see also Panksepp, J. & Biven, L. *Archaeology of Mind: Neuroevolutionary Origins of Human Emotions*. (W. W. Norton, 2012).
17. Penfield & Jasper, 1954; see also Blumenfeld, H. 'Consciousness and epilepsy: why are patients with absence seizures absent?' *Progress in Brain Research*, 150, 271–86 (2005).
18. A recent review by Anthony Hudetz highlights global reductions in cerebral metabolism during general anaesthesia. He does comment that there is regional heterogeneity but such heterogeneity is different depending on which anaesthetic is used, suggesting there is no common, specific brain area responsible. Networks are likely to be more important. He states: 'Anesthetic loss of consciousness is not a block of corticofugal information transfer, but a disruption of higher-order cortical information integration. The prime candidates for functional networks of the forebrain that play a critical role in maintaining the state of consciousness are those based on the posterior parietal-cingulate-precuneus region and the nonspecific thalamus.' Hudetz, A. G. 'General anesthesia and human brain connectivity'. *Brain Connectivity*, 2, 291–302 (2012).

19. Although widely cited, the only reference to be found is in a textbook, *Psychology in Medicine*, published in 1992 (now out of print), by I. Chris McManus. In Chapter 23 he writes: 'Remember the old anecdote of removing a valve from a radio set, hearing it start to whistle, and falsely assuming the valve's function is to stop the set whistling. Pathological functioning can only be interpreted through an adequate knowledge of normal functioning.' http://www.ucl.ac.uk/medical-education/publications/psychology-in-medicine
20. Yet only recently has the prospect of a privileged brain area for consciousness once again resurfaced. Mohamad Koubeissi and his colleagues took recordings of the brain of a severely epileptic patient who was, because of his condition, already undergoing invasive neurosurgery. By sheer happenstance, they discovered that stimulation of a particular area (the claustrum) reversibly rendered the patient unconscious: once the stimulation was terminated, then the individual was once again fully responsive. Had the scientists found, if not the 'centre for' consciousness, then at least the control centre for it, as originally suggested some time ago by Francis Crick? Unsurprisingly, the answer has to be no. There are a variety of problems with this report. The first are simply technical: the finding is hardly based on an appropriate sample size but on just one patient, and moreover a patient with an abnormal brain that was missing part of a region (hippocampus) related to memory; any convincing evidence would have to be derived from a far larger group of subjects. In addition, since completely healthy individuals would never be used for invasive brain surgery, the degree of brain damage would have to be consistent within the cohort and any conclusions reached would therefore be qualified by the limitation that this brain damage could be making an all-important difference. The second type of problem lies with the danger of overinterpretation. While the claustrum might be the only area currently documented which, when stimulated, induces unconsciousness, how can we say for certain that this is an exclusive effect until every single other brain area has been exhaustively stimulated under similar conditions? In fact, neuroscientists have known for several years that stimulation of completely different and multiple areas (loci between the internal globus pallidus and nucleus basalis of Meynert) can suddenly restore consciousness in patients who are otherwise anaesthetized. So what makes the claustrum unique as a brain region? At most, and leaving aside all technical reservations, the best, surely, that we can say is that it is connected to many other brain areas – the very reason Crick favoured it in the first place. Stimulation of this area, as indeed with areas such as the globus

pallidus, would then in some way lead to changes in global brain states. The third type of problem revolves around how we define consciousness, in this case, by using the operational strategy. It turns out that the patient in Koubeissi's study was not unconscious in the same way as they would have been if anaesthetized or asleep: rather, the stimulation resulted in 'a complete arrest of volitional behavior and unresponsiveness', perhaps more analogous to being in some kind of trance. But does that mean that the patient was really unconscious in a clinical sense, or that it could be guaranteed they were having no subjective experience? Bizarrely, Anil Seth from Sussex University interpreted this finding by attempting to differentiate between consciousness and wakefulness: apparently, the patient was awake but not conscious! Yet the dictionary, for what it's worth, effectively equates the two and defines consciousness as 'full activity of the mind and senses, as in waking life'. Such nuances in definition as employed by Seth, unless elaborated and justified much more fully, surely defy all common sense. The fourth type of problem relates to the over-reliance on metaphor in order to understand the data. Francis Crick, whose ideas are allegedly validated by this report, had likened the claustrum to the 'conductor in an orchestra', coordinating all the other brain areas. But what would that actually mean? That the claustrum is a kind of Boss? If it's simply because this brain area is at a kind of anatomical crossroads, then why should it also automatically have executive functions, rather than simply being some kind of coordinator, at best, or a conduit, at worst? In any event, Koubeissi chose a completely different metaphor to Crick and likened the stimulation of the claustrum to a switch – say, the ignition key of a car – that turns consciousness on and off. Yet surely a switch and an orchestral conductor have very different functions. Moreover, even in the same report, the 'switch' appears to have been more analogue than simply on–off and not a simple switch at all: '... she [the patient] gradually spoke more quietly or moved less and less until she drifted into unconsciousness'. So the very assumptions that the researchers were observing the induction of unconsciousness need to be challenged, along with the idea that consciousness was then simply 'switched' on and off. Koubeissi, M. Z. et al. 'Electrical stimulation of a small brain area reversibly disrupts consciousness'. *Epilepsy and Behavior*, 37, 32–5 (2014). Stevens, C. F. 'Consciousness: Crick and the claustrum'. *Nature*, 435, 1040–1 (2005). Bagary, M. 'Epilepsy, consciousness and neurostimulation'. *Behavioural Neurology*, 24, 75–81 (2011).

21. Haynes, J. & Rees, G. 'Decoding mental states from brain activity in humans'. *Nature Reviews Neuroscience* 7, 523–34 (2006).

22. Blakemore, C. & Greenfield, S. A., Hacker Peter eds., Ch. 31 in *Mindwaves: Thoughts on Intelligence, Identity and Consciousness*, 485–505 (1987).
23. Maguire, E. A. et al. 'Navigation-related structural change in the hippocampi of taxi drivers'. *Proceedings of the National Academy of Sciences of the United States of America*, 97, 4398–403 (2000).
24. García-Lázaro, H. G. et al. 'Neuroanatomy of episodic and semantic memory in humans: a brief review of neuroimaging studies'. *Neurology India*, 60, 613–17; see also Squire, L. R. 'Memory and brain systems: 1969–2009'. *Journal of Neuroscience*, 29, 12711–16 (2009).
25. Alkire, M. T. et al. 'Thalamic microinjection of nicotine reverses sevoflurane-induced loss of righting reflex in the rat'. *Anesthesiology*, 107, 264–72 (2007).
26. Posner, J. B. & Plum, F. *Plum and Posner's Diagnosis of Stupor and Coma*. (Oxford University Press, 2007); see also Miller, J. W. & Ferrendelli, J. A. 'The central medial nucleus: thalamic site of seizure regulation'. *Brain Research*, 508, 297–300 (1990); and Miller, J. W. & Ferrendelli, J. A. 'Characterization of GABAergic seizure regulation in the midline thalamus'. *Neuropharmacology*, 29, 649–55 (1990).
27. Alkire, M. T., Haier, R. J. & Fallon, J. H. 'Toward a unified theory of narcosis: brain imaging evidence for a thalamocortical switch as the neurophysiologic basis of anesthetic-induced unconsciousness'. *Consciousness and Cognition*, 9, 370–86 (2000).
28. Tononi, G. 'An information integration theory of consciousness'. *BMC Neuroscience*, 5, 42–64 (2004); see also Massimini, M. et al. 'Triggering sleep slow waves by transcranial magnetic stimulation'. *Proceedings of the National Academy of Sciences of the United States of America*, 104, 8496–501 (2007).
29. Gamma oscillations specifically appear to depend on recurrent connections between the excitatory neurons and a subset of the remaining inhibitory brain cells which are able to discharge rapidly and predominantly from synapses close to the main part (cell body) of the excitatory neurons: the whole point of this set-up would be that during gamma activity each excitatory cell need fire action potentials only on a small fraction of cycles, but the total contributing to each cycle provides enough excitatory drive to the inhibitory neurons to make them fire on every cycle: a kind of neuronal Marxism, in which each individual working intermittently nonetheless ensures a continuous output collectively. Mann, E. O. et al. 'Perisomatic feedback inhibition underlies cholinergically induced fast network oscillations in the rat hippocam-

pus in vitro'. *Neuron*, 45, 105–17 (2005). Fisahn, A. et al. 'Cholinergic induction of network oscillations at 40Hz in the hippocampus in vitro'. *Nature*, 394, 186–9 (1998).

30. Singer, W. 'Neuronal synchrony: a versatile code for the definition of relations?' *Neuron*, 24, 49–65, 111–25 (1999); see also Singer, W. & Gray, C. M. 'Visual feature integration and the temporal correlation hypothesis'. *Annual Review of Neuroscience*, 18, 555–86 (1995); and Tononi, G., Sporns, O. & Edelman, G. M. 'Re-entry and the problem of integrating multiple cortical areas: simulation of dynamic integration in the visual system'. *Cerebral Cortex*, 2, 310–35 (1992).
31. Wu, J. Y. et al. 'Spatiotemporal properties of an evoked population activity in rat sensory cortical slices'. *Journal of Neurophysiology*, 86, 2461–74 (2001).
32. Koubeissi et al., 2014.
33. Tononi, G. & Koch, C. 'The neural correlates of consciousness: an update'. *Annals of the New York Academy of Sciences*, 1124, 239–61 (2008).
34. Bachmann, T. *Microgenetic Approach to the Conscious Mind*. (John Benjamins, 2000).
35. Sergent, C., Baillet, S. & Dehaene, S. 'Timing of the brain events underlying access to consciousness during the attentional blink'. *Nature Neuroscience*, 8, 1391–400 (2005).
36. Libet, B. *Mind Time: The Temporal Factor in Consciousness*. (Harvard University Press, 2004).
37. Edelman, G. M. *The Remembered Present: A Biological Theory of Consciousness*. (Basic Books, 1989).
38. Rossi, A. F., Desimone, R. & Ungerleider, L. G. 'Contextual modulation in primary visual cortex of macaques'. *Journal of Neuroscience*, 21, 1698–709 (2001).
39. Pascual-Leone, A. & Walsh, V. 'Fast back projections from the motion to the primary visual area necessary for visual awareness'. *Science*, 292, 510–12 (2001).
40. Todman, D., Wilder Penfield (1891–1976)'. *Journal of Neurology*, 255, 1104–5 (2008).
41. Quiroga, R. et al. 'Invariant visual representation by single neurons in the human brain'. *Nature*, 435, 1102–7 (2005).
42. Kandel, E. R. 'An introduction to the work of David Hubel and Torsten Wiesel'. *Journal of Physiology*, 587, 2733–41 (2009).
43. Gross, C. G. 'Genealogy of the "grandmother cell"'. *Neuroscientist*, 8, 512–18 (2002); see also Jagadeesh, B. 'Recognizing grandmother'. *Nature Neuroscience*, 12, 1083–5 (2009).

44. Quiroga, R. Q., Fried, I. & Koch, C. 'Brain cells for grandmother.' *Scientific American*, 308, 30–5 (2013).
45. Biederman, I. 'Recognition-by-components: a theory of human image understanding'. *Psychological Review*, 94, 115–47 (1987); see also Marr, D. & Nishihara, H. K. 'Representation and recognition of the spatial organization of three-dimensional shapes'. *Proceedings of the Royal Society of London B: Biological Sciences*, 200, 269–94 (1978); and Booth, M. C. A. & Rolls, E. T. 'View-invariant representations of familiar objects by neurons in the inferior temporal visual cortex'. *Cerebral Cortex*, 8, 510–23 (1998); and Bulthoff, H. H., Edelman, S. Y. & Tarr, M. J. 'How are three-dimensional objects represented in the brain?' *Cerebral Cortex*, 5, 247–60 (1995); and Vetter, T., Hurlbert, A. & Poggio, T. 'View-based models of 3D object recognition: invariance to imaging transformations'. *Cerebral Cortex*, 5, 261–9 (1995); and Logothetis, N. K. & Pauls, J. 'Psychophysical and physiological evidence for viewer-centered object representations in the primate'. *Cerebral Cortex*, 5, 270–88 (1995).
46. Connor, C. E. 'Neuroscience: friends and grandmothers'. *Nature*, 435, 1036–7 (2005).
47. Their idea was that the principles of quantum theory (the physics of the very small) could provide an alternative process for the generation of consciousness that would involve various hypothetical possibilities being reduced by empirical observation to one single certainty (collapse of the wave function). In one version of quantum theory (the Copenhagen Interpretation, which attempts to reconcile the theoretical formulations of quantum mechanics to corresponding experimental data), the act of observation itself causes the system being watched eventually to be in one type of erstwhile various potential states – 'subjective reduction' – since an observer is required. But in the brain, since there is no outside observer, quantum events would have to 'collapse' spontaneously, without anyone watching: that is, objectively. Cramer, J. 'The transactional interpretation of quantum mechanics'. *Review of Modern Physics*, 58, 647–87 (1986).
48. Hameroff, S. & Penrose, R. 'Consciousness in the universe: a review of the "Orch OR" theory'. *Physics of Life Reviews*, 11, 39–78 (2014).
49. Inevitably, there are many questions and problems raised by this highly original scenario. The most immediate is that the brain is much too hot for quantum events to occur, since their coherence requires an ambient temperature of 310° Kelvin, and this would be disturbed by the thermal energy of the naturally hot brain. On the other hand, the physicist Herbert Fröhlich has suggested that the brain might be a very special

exception, where, unlike events in the outside world, coherence in the brain could be generated and sustained by energy provided by chemical reactions within the cell. He estimated that the tubulin molecules could be excited in a coherent fashion for between 10^{-12} to 10^{-9} seconds, that is, a nanosecond. However, if a moment of consciousness is about 500msecs, as suggested by the work of Benjamin Libet (2004), then approximately 10^{9} tubulin molecules would be needed, namely the number found in a total of some 100 to 100,000 neurons. But this number would be too few to achieve meaningful quantum coherence: even a simple and brief flash of light will activate an estimated 10^{7} neurons in the brain of a cat. However, more recently, Hameroff and Penrose have responded to the seemingly deal-breaking incompatibility of quantum events with brain heat by citing observations that certain phenomena (quantum spin transfer) are enhanced at increasingly warm temperatures, as well as pointing out that plants use quantum events in photosynthesis at ambient temperatures. Warm quantum effects also occur in bird-brain navigation, ion channels (by which the pores in cell membranes allow ions such as sodium and potassium to pass from exterior to interior, and vice versa), smell, and protein folding (the process by which a protein chain aquires its three-dimensional structure). A revised description of consciousness founded on quantum theory published in 2013 is now based on 'beat frequencies' of microtubules allegedly corresponding to brainwave oscillations at frequencies such as 40Hz, which, as we've seen, has enjoyed an enduring popularity as a neural correlate of consciousness: but we've also seen that such a collective synchrony is far from accepted as the gold standard. In fact, it might well be that steady-state oscillations are the default mode of the brain, the background 'noise' against which the all-important salient 'signal' of a conscious experience is superimposed. If so, of course the million-dollar question would be to identify and describe that signal. More significantly, however, for the theory, an energy supply has been identified within cells from a specific biochemical reaction (hydrolysis of guanosine triphosphate to guanosine diphosphate). At least a part of the energy supplied from this breakdown can excite vibrations, thereby attributable to microtubule dynamics, driven through the electromagnetic fields from the powerhouse of the cell (mitochondrion). Bernroider, G. & Roy, S. 'Quantum entanglement of K+ ions, multiple channel states and the role of noise in the brain'. SPIE Third International Symposium on Fluctuations and Noise (eds. Stocks, N. G., Abbott, D. & Morse, R. P.), 205–14 (International Society for Optics and Photonics, 2005). Engel, G. S. et al.

'Evidence for wavelike energy transfer through quantum coherence in photosynthetic systems'. *Nature*, 446, 782–6 (2007). Fröhlich, H. 'The extraordinary dielectric properties of biological materials and the action of enzymes'. *Proceedings of the Natural Academy of Sciences of the United States of America*, 72, 4211–15 (1975). Grinvald, A. et al. 'Cortical point-spread function and long-range lateral interactions revealed by real-time optical imaging of macaque monkey primary visual cortex'. *Journal of Neuroscience*, 14, 2545–68 (1994). Gauger, E. M. et al. 'Sustained quantum coherence and entanglement in the avian compass'. *Physical Review Letters*, 106, 040503 (2011). Hildner, R. et al. 'Quantum coherent energy transfer over varying pathways in single light-harvesting complexes'. *Science*, 340, 1448–51 (2013). Libet, B., Wright, E. W. & Gleason, C. A. 'Preparation- or intention-to-act in relation to pre-event potentials recorded at the vertex'. *Electroencephalography and Clinical Neurophysiology*, 56, 367–72 (1983). Ouyang, M. & Awschalom, D. D. 'Coherent spin transfer between molecularly bridged quantum dots'. *Science*, 301, 1074–8 (2003). Pokorný, J. 'Excitation of vibrations in microtubules in living cells'. *Bioelectrochemistry*, 63, 321–6 (2004). Turin, L. 'A spectroscopic mechanism for primary olfactory reception'. *Chemical Senses*, 21, 773–91 (1996).

50. Hameroff and Penrose emphasize that neuronal cohesion could be achieved by the spread of activity via 'gap junctions' (low-resistance electrical contacts where one cell is continuous with another): but the Orch OR scheme would need to cater for the well-known chemical diversity that can modify consciousness so powerfully, via various psychoactive drugs. In any case, there is no reason just yet to give up on classic synaptic signalling. One way of ensuring that more neurons could be involved, that the brain microtubules behave differently and that specific chemical messengers could play a varied part comes from the following suggestion from the neuroscientist Nancy Woolf. Before tubulin within microtubules can reconfigure so that the appropriate quantum waves can be generated, a particular protein (microtubule associated protein, MAP2) has to be put out of action: this chemical is a bit like glue in keeping microtubules configured in an independent, local pattern. Interestingly enough, MAP2 is not generic to all cells, and even within the cortex in the brain is located only in some 15 per cent of cells: moreover, it is activated only by certain chemical messengers binding to their receptors. These are the exact requirements that neuroscientists need for unpacking consciousness in terms of different brain regions, drug sensitivity, and so on. When a transmitter binds to

its molecular target (receptor), it will inhibit MAP2, thereby enabling the microtubules to aggregate into the rows of uniform, parallel alignment necessary for a flash flood of quantum 'decoherence' and a moment of consciousness. As soon as the transmitter action is completed, the microtubules will reconfigure into a new pattern. The brain will have changed for ever, and that moment of consciousness will never come again. This scenario could be attractive in unifying quantum theory with macro-scale events. However, the problem with this within-cell NCC – be it the original version from Penrose and Hameroff, its revised version or the hybrid variant envisaged here – is that it nonetheless rests on many assumptions and nothing less than a new application of physics. It is also currently without any empirical validation whatsoever. Woolf, N. J. 'A possible role for cholinergic neurons of the basal forebrain and pontomesencephalon in consciousness'. *Consciousness and Cognition*, 6, 574–96 (1997).

51. Dehaene, S., Kerszberg, M. & Changeux, J. P. 'A neuronal model of a global workspace in effortful cognitive tasks'. *Proceedings of the Natural Academy of Sciences of the United States of America*, 95, 14529–34 (1998); see also Barrs, B. J. *A Cognitive Theory of Consciousness*. (Cambridge University Press, 1988).
52. Dennett, D. C. 'Are we explaining consciousness yet?' *Cognition*, 79, 221–37 (2001); see also Dennett, D. C. *Consciousness Explained*. (Basic Books, 1991).
53. Panpsychism literally means that everything ('pan') has a mind ('psyche'): for a modern account of this ancient line of thought, see the writings of David Chalmers, for example, http://consc.net/papers/panpsychism.pdf
54. Tononi, 2004. See also E. Tononi, 'Integrated information theory of consciousness: an updated account'. *Archives Italiennes Biologie*, 150, 290–326 (2012).
55. Tononi introduces the term 'phi' as a measure of this integrated information: local collections of neurons in key brain areas will maximize phi over a time window of ten to hundreds of milliseconds. So the thalamocortical system, with all the reciprocal complexity it exhibits, is cited as an example of 'high' phi.
56. McGinn, C. *The Mysterious Flame: Conscious Minds in a Material World*. (Basic Books, 1999).
57. Crick, F. & Koch, (2003)
58. Dennett, 1991.
59. Libet, Wright & Gleason, 1983.
60. Tononi, Sporns & Edelman, 1992.

61. Kurzweil, R. *How to Create a Mind: The Secret of Human Thought Revealed*. (Viking, 2012).
62. https://gigaom.com/2014/06/25/googles-ray-kurzweil-on-the-moment-when-computers-will-become-conscious/
63. Bergquist, F. & Ludwig, M. 'Dendritic transmitter release: a comparison of two model systems'. *Journal of Neuroendocrinology*, 20, 677–86 (2008).
64. Damasio, A. *The Feeling of What Happens: Body, Emotion and the Making of Consciousness*. (Harcourt Brace, 2000).
65. Anil Seth, from the University of Sussex, has distinguished models of consciousness from NCCs and, indeed, 'theories' in general: the key distinction, he claims, is that models alone provide 'an explanatory link' between neural activity and consciousness. However, although he has suggested that the crucial difference for the models he so favours is 'mechanistic implementation', few such models to date are truly based on novel neuronal mechanisms in which there is good reason to think that they will be not just necessary but also sufficient for consciousness. In contrast to Seth's definition, I would argue that at an even more fundamental level the crucial feature of a model is surely that it extracts and reproduces the salient features in a system at the expense of the extraneous ones. For example, if you wanted to model flight, the crucial feature would be to defy gravity, and you could jettison any consideration for beaks and feathers. But when you want to 'model' consciousness, how do you know in advance what the salient feature is? And surely if you did know it, then you wouldn't really need to construct a model in the first place. Seth, A. 'Models of consciousness'. *Scholarpedia*, 2, 1328 (2007).
66. McGinn, C. *The Mysterious Flame: Conscious Minds in a Material World*. (Basic Books, 1999).
67. Chalmers, D. J. 'The puzzle of conscious experience'. *Scientific American*, 273, 80–6 (1995); see also Chalmers, D. J. 'Facing up to the problem of consciousness'. *Journal of Consciousness Studies*, 2, 200–19 (1995).

2. WAKING UP

1. Theta waves have a characteristic amplitude of 10 microvolts and a frequency of four to eight cycles per second. Lancel, M. 'Cortical and subcortical EEG in relation to sleep–wake behavior in mammalian species'. *Neuropsychobiology*, 28(3), 154–9 (1993). Başar E. &

Güntekin B. 'Review of delta, theta, alpha, beta and gamma response oscillations in neuropsychiatric disorders'. *Supplements to Clinical Neurophysiology*, 62, 303–41 (2013).

2. The duration of REM sleep increases from ten minutes in the first cycle up to fifty minutes in the final cycle. See Chapter 28 and Figure 28.7A in Purves, D. et al. (eds.) *Neuroscience*. (Sinauer, 2012).
3. Ibid.
4. Dopamine, noradrenaline, histamine, serotonin and acetylcholine have long been established as 'classic' transmitters, released from the end of a specialised brain-cell process (the axon terminal) to cross the narrow gap (synapse) on to the next target cell: the transmitter then enters into a molecular handshake with its own specialized, bespoke protein (a receptor) such that a new action potential is generated in the target cell. For a detailed account see Kandel, E., Schwartz, James H. & Jessell, T., *Principles of Neural Science*, 5th edn (Elsevier, 2012).
5. Aston-Jones, G. & Bloom, F. E. 'Activity of norepinephrine-containing locus coeruleus neurons in behaving rats anticipates fluctuations in the sleep-waking cycle'. *Journal of Neuroscience*, 1, 876–86 (1981); see also Kocsis, B. et al. 'Serotonergic neuron diversity: identification of raphe neurons with discharges time-locked to the hippocampal theta rhythm'. *Proceedings of the Natural Academy of Sciences of the United States of America*, 103, 1059–64 (2006); and Steininger, T. L. et al. 'Sleep–waking discharge of neurons in the posterior lateral hypothalamus of the albino rat'. *Brain Research*, 840, 138–47 (1999); and Takahashi, K., Lin, J.-S. & Sakai, K. 'Neuronal activity of histaminergic tuberomammillary neurons during wake–sleep states in the mouse'. *Journal of Neuroscience*, 26, 10292–8 (2006); and Takahashi, K. et al. 'Locus coeruleus neuronal activity during the sleep–waking cycle in mice'. *Neuroscience*, 169, 1115–26 (2010); and Jacobs, B. L. & Fornal, C. A. 'Activity of brain serotonergic neurons in the behaving animal'. *Pharmacological Reviews*, 43, 563–78 (1991).
6. Hobson, J. A. 'Sleep and dreaming: induction and mediation of REM sleep by cholinergic mechanisms'. *Opinion in Neurobiology*, 2, 6, 759–63 (Dec. 1992).
7. Lee, S. H. & Dan, Y. 'Neuromodulation of brain states'. *Neuron*, 76, 109–222 (2012).
8. Greenfield, S. A. *The Private Life of the Brain* (Penguin, 2000). For two contrasting examples of modulation by acetylcholine, see: Cole, A. E. & Nicoll, R. A. 'Acetylcholine mediates a slow synaptic potential in hippocampal pyramidal cells'. *Science*, 221, 1299–301 (1983); and

McCormick, D. A. & Prince, D. A. 'Mechanisms of action of acetylcholine in the guinea-pig cerebral cortex in vitro'. *Journal of Physiology*, 375, 169–94 (1986).

9. Guedel, A. E. *Inhalational Anesthesia: A Fundamental Guide*. (Macmillan, 1937).
10. Meanwhile, a possible fourth stage – which is only a theoretical state – is one in which so much anaesthetic has been administered that the brain cells in the primitive part of the brain just above the spinal cord (the brainstem) remain alive but stop firing their all-important action potentials. Since these cells control respiration and heart rate, you would therefore stop breathing and your blood pressure would fall dangerously low, impairing perfusion of vital organs. In other words, you would be dead. But this isn't too great a cause for concern: since these 'dangers' Guedel described were probably due to a combination of hypoxia and hypotension, which are indeed caused by high-dose anaesthetic effects on the brain, these dangers can be mitigated nowadays and are potentially reversible.
11. BIS monitoring measures the effect of anaesthetic on the magnitude and/or coordination of brain electrical signals, their outputs being represented as a single final number. For this reason, the real value of BIS and the information it actually conveys are debated: however, all clinicians acknowledge that one of the drawbacks is that BIS measurements do not reflect all anaesthetic action equally and are insensitive to certain agents that nonetheless induce unconsciousness (for example, nitrous oxide, ketamine and xenon). One theory for this restricted sensitivity is that the BIS EEG-based monitors track specific brain changes induced only by certain anaesthetics acting predominantly via some chemical systems but not others. Pandit, J. J. & Cook, T. M. 'National Institute for Clinical Excellence guidance on measuring depth of anaesthesia: limitations of EEG-based technology'. *British Journal of Anaesthesia*, 112, 385–6 (2014).
12. Sometimes conclusions are greatly dependent upon statistical interpretations which are not self-evident and require great effort and training to replicate. For example, UnCheol Lee and his team reported that ketamine did indeed produce different effects on the EEG to those produced by propofol and sevoflurane. To reach this conclusion, they had to use complex mathematical and statistical algorithms; but, at the same time, they acknowledged that other studies using similarly complex analytical functions had produced generally opposite results. Barrett, A. B. et al. 'Granger causality analysis of steady-state electroencephalographic signals during propofol-induced anaesthesia'. *PLoS*

General overview of anaesthetic actions on the main putative candidate receptors/channels

	$GABA_A$	Glycine	nACh (muscle)	nACh (neuronal)	$5HT_3$	AMPA	Kainate	NMDA	TASK1	HCN1
Etomidate					0	?	?	?		0
Propofol					0		0		0	
Barbiturates								0	?	?
Ketamine		0				0	0		?	
Isoflurane										
Sevoflurane					?	0	?			?
Nitrous oxide									?	?
Dexmedetomidine	0	?	0	0	0	?	?	?	?	?

Key:	Strongly activating	Weakly activating	Weakly inhibiting	Strongly inhibiting	0 = no effect	? = unknown

GABA-A, γ-amino butyric acid type A; nACh, nicotinic acetylcholine; 5HT3, 5-hydroxytryptamine (serotonin) type 3; AMPA, α-amino-3-hydroxy-5-methyl-4-isoxazolepropionic acid; NMDA, N-methyl-D-aspartate; TASK-1, TWIK (two-pore domain weakly inwardly rectifying potassium)-related, acid sensitive potassium channel type 1; HCN1, hyper-polarization-activated cation channel type 1.

Adapted from Rudolph, U. & Antkowiak, B. 'Molecular and neuronal substrates for general anaesthetics'. *Nature Reviews Neuroscience*, 709–20 (2004).

One, 7 (2012). Cruse, D. et al. 'Detecting awareness in the vegetative state: electroencephalographic evidence for attempted movements to command'. *PLoS One*, 7, e49933 (2012). Goldfine, A. M. et al. 'Reanalysis of "Bedside detection of awareness in the vegetative state: a cohort study"'. *Lancet*, 381, 289–91 (2013). Lee, U. et al. 'Disruption of frontal-parietal communication by ketamine, propofol and sevoflurane'. *Anesthesiology*, 118, 1264–75 (2013). Mashour, G. A. & Avidan, M. S. 'Capturing covert consciousness'. *Lancet*, 381, 271–2 (2013). Menon, R. & Kim, S. 'Spatial and temporal limits in cognitive neuroimaging with fMRI'. *Trends in Cognitive Science*, 3, 207–16 (1999). Nicolaou, N., Hourris, S., Alexandrou, P. & Georgiou, J. 'EEG-based automatic classification of "awake" versus "anesthetized" state in general anesthesia using Granger causality'. *PLoS One*, 7, e33869 (2012). A recent attempt by one team using complex EEG analysis argued that patients in a persistent vegetative state were likely to be aware: but this suggestion was refuted with re-analysis of the same data by other groups. In fact, yet another clinic went so far as to claim: 'Interpretation of states of consciousness exclusively on the basis of neurophysiological data is crucially dependent on the statistical model.' Meanwhile, another method, brain imaging, is far from ideal. fMRI and other scanning techniques will give a read-out of blood flow within the brain but not directly of neuronal activity: the time window for measuring brain events, as we saw in the previous chapter, is not therefore commensurate with real-time effects. So, with these methods, discrepancies between different anaesthetics is hard to evaluate, and therefore even harder to explain in terms of a single, common, final brain process.

13. Stiles, J. & Jernigan, T. L. 'The basics of brain development'. *Neuropsychology Review*, 20, 327–48 (2010).
14. Greenfield, S. A. *Journey to the Centres of the Mind*. (W. H. Freeman, 1995); Greenfield, S. A. *The Private Life of the Brain: Emotions, Consciousness and the Secret of the Self*. (Wiley, 2000). Koch, C. & Greenfield, Susan. 'How does consciousness happen?' *Scientific American*, 297, 76–83 (2007).
15. Tononi, G. & Koch, C. 'The neural correlates of consciousness: an update'. *Annals of the New York Academy of Sciences*, 1124, 239–61 (2008).
16. For instance (using halothane anaesthesia), frontal, temporal, parietal, occipital lobes, anterior cingulate, basal ganglia, thalamus, hippocampus, midbrain and cerebellum all showed reduced glucose metabolism during anaesthesia. Alkire, M. T. et al. 'Functional brain imaging dur-

ing anesthesia in humans: effects of halothane on global and regional cerebral glucose metabolism'. *Anesthesiology*, 90, 701–9 (1999).

17. Alkire, M. T., Hudetz, A. G. & Tononi, G. 'Consciousness and anesthesia'. *Science*, 322, 876–80 (2008).
18. Lewis, L. D. et al. 'Rapid fragmentation of neuronal networks at the onset of propofol-induced unconsciousness'. *Proceedings of the Natural Academy of Sciences of the United States of America*, 109, E3377–86 (2012).
19. Massimini, M. et al. 'Triggering sleep slow waves by transcranial magnetic stimulation'. *Proceedings of the Natural Academy of Sciences of the United States of America*, 104, 8496–501 (2007).
20. Hebb, D. O. *The Organization of Behavior: A Neuropsychological Theory*. (Wiley, 1949).
21. Spatz, H. C. 'Hebb's concept of synaptic plasticity and neuronal cell assemblies'. *Behavioural Brain Research*, 78, 3–7 (1996).
22. Kandel, E. R. & Schwartz, J. H. 'Molecular biology of learning: modulation of transmitter release'. *Science*, 218, 433–43 (1982).
23. Hebb, 1949.
24. Voltage-sensitive dye imaging (VSDI) has evolved from being a novel and rarified technique developed over twenty years ago into a highly successful and widely used means of representing neural activity. In 1968, Tasaki et al. were the first to use VSDI to measure electric activity in the squid giant axon. The technology was developed primarily by Alan Waggoner (Waggoner, A. S. *Journal of Membrane Biology*, 27, 317–34, who synthesized and screened a large number of various dyes. The origin of action in the dye used by us is electrochromism, more generally known as the Stark effect, which is the electric equivalent to the magnetic Zeeman effect. Three criteria, while not sufficient, are required for a dye to be categorized as electrochromic: 1) electrochromic shifts occur on a sub-nanosecond timescale, as they do not rely on molecular movement, 2) the first derivative of the absorption or excitation spectrum should have the same slope as the difference spectrum as $\Delta\varepsilon$ is proportional to $\partial\varepsilon/\partial\lambda$ (ε = 'effective' extinction) and 3) the chromophore should be asymmetric. While these criteria are not met by the early merocyanine, cyanine and oxonol dyes, di-4-ANEPPS, as well as other newer dyes, do meet these criteria. These dyes all contain the aminostyryl pyridinium chromophore. According to the theory, photoexcitation results in a shift of the positive charge away from the pyrimidinium end towards the aminophenyl end of the molecule. The chromophore is anchored in the membrane so that the vector of this shift is perpendicular to the membrane plane and thus parallel to the electric field. The shift is estimated to be of the order of 0.3 nm (where

a nanometre (nm) is a billionth of a metre). It is believed that the electric field thus hinders the intramolecular relocation of the positive charge, resulting in the voltage-sensitive optical effects measured. For di-4-ANEPPS, a further mechanism must be taken into account. A field-induced 'resolvation' moves the charged naphthalene group towards the surface of the membrane and twists the molecule, resulting in a weaker blueshift in the fluorescence spectrum, a decreased quantum yield and a broadening of the spectrum. Grinvald, A. et al. 'Cortical point-spread function and long-range lateral interactions revealed by real-time optical imaging of macaque monkey primary visual cortex'. *Journal of Neuroscience*, 14, 2545–68 (1994). Cohen, L. B. et al. 'Changes in axon fluorescence during activity: molecular probes of membrane potential'. *Journal of Membrane Biology*, 19, 1–36 (1974). Waggoner, A. S. & Grinvald, A. 'Mechanisms of rapid optical changes of potential sensitive dyes'. *Annals of the New York Academy of Sciences*, 303, 217–41 (1977). Waggoner, A. S. 'The use of cyanine dyes for the determination of membrane potentials in cells, organelles and vesicles'. *Methods in Enzymology*, 55, 689–95 (1979). Fluhler, E., Burnham, V. G. & Loew, L. M. 'Spectra, membrane binding and potentiometric responses of new charge shift probes'. *Biochemistry*, 24, 5749–55 (1985). Ebner, T. J. & Chen, G. 'Use of voltage-sensitive dyes and optical recordings in the central nervous system'. *Progress in Neurobiology*, 46, 463–506 (1995). Fromherz, P. & Lambacher, A. 'Spectra of voltage-sensitive fluorescence of styryl-dye in neuron membrane'. *Biochimica et Biophysica Acta*, 1068, 149–56 (1991).

25. However, VSDI itself cannot detect single action potentials: the spatial resolution is some 100x100x100 micrometres, which would encompass about fifty to a hundred neuron cell bodies along with their connective processes to and from hundreds of other neurons. So a first crucial step in quantifying and validating these highly transient coalitions has been pioneered by Amiram Grinvald and his group, who showed that spontaneous firing of single neurons is correlated to the activity of their associated neuronal assembly. The VSDI technique, while needing complementary techniques for the bottom-up approach, is, on the other hand, perfectly fitted for revealing the meso-scale level of brain organization. See also Grinvald, A. et al. 'Cortical point-spread function and long-range lateral interactions revealed by real-time optical imaging of macaque monkey primary visual cortex'. *Journal of Neuroscience*, 14, 2545–68 (1994). Arieli, A. & Grinvald, A. 'Optical imaging combined with targeted electrical recordings, microstimulation or tracer injections'. *Journal of Neuroscien-*

tific Methods, 116, 15–28 (2002). Tominaga, T. et al. 'Quantification of optical signals with electrophysiological signals in neural activities of di-4-ANEPPS stained rat hippocampal slices'. *Journal of Neuroscientific Methods*, 102, 11–23 (2000).

26. This definition of assemblies is not universally adopted: the term has been used for very different phenomena, for example, as synonymous with cortical columns that are 'anatomically well-defined examples of neuronal networks', with sleep seen as merely a product of previous activity states, in turn dependent on conventional transmission. Krueger J. M. et al. 'Sleep as a fundamental property of neuronal assemblies'. *National Review of Neuroscience*, 9(12): 910–19 (2008).
27. Grinvald et al. 1994.
28. Devonshire, I. M. et al. 'Effects of urethane anaesthesia on sensory processing in the rat barrel cortex revealed by combined optical imaging and electrophysiology'. *European Journal of Neuroscience*, 32, 786–97 (2010).
29. Grinvald et al., 1994.
30. If you assume the average diameter of a brain cell is about 40 um (about forty thousandths of a metre) and that between 5 and 10 per cent of total brain volume is the fluid that bathes those cells, then in this particular instance, where the activity is maximal at about 10msec, then this would be some 1,265 neurons in diameter, and if the spread of this activity were a perfect sphere, 1.06 billion neurons. Of course, the three-dimensional shape is not possible to appreciate with this type of experiment on a two-dimensional slice and, even if it were, then there would be no reason to assume that the coalition would be spherical: but the main point is that you can get an idea of the large number of neurons involved. The downside of VSDI is that it cannot be used in human brains, as it is invasive, with the potentially toxic dye working directly with the living brain cells. Nonetheless, the non-invasive techniques used in human brain imaging can offer only highly indirect measures of brain activity, namely the increases in the blood supply required by more active brain cells: as such, there will always have to be a delay, so the response is averaged over several seconds. Although some advances are now being made to develop real-time non-invasive imaging, the resolution is still several orders of magnitude longer than an action potential. Stoeckel, L. E. et al. 'Optimizing real-time fMRI neurofeedback for therapeutic discovery and development'. *NeuroImage: Clinical*, 5, 245–55 (2014). So, for example, the sequence of events recorded in Figure 2 using VSDI would have still been entirely missed.

An important difference, however, is that the assembly triggered in the intact sensory cortex is even bigger than *in vitro*, measuring at least 2.5mm across rather than 1mm, as in the experiments with slices: this discrepancy is presumably because the whole event is more 'true to life' where there is a greater availability of naturally occurring modulators, more connectivity and an even greater number of working inputs into the brain. Devonshire et al., 2010.

31. Llinás, R. & Sasaki, K. 'The functional organization of the olivo-cerebellar system as examined by multiple purkinje cell recordings. *European Journal of Neuroscience*, 1, 587–602 (1989).

32. Wu has suggested four further mechanisms for this unevoked wave propagation: firstly, a single oscillator excites with different time delays; secondly, direct excitement from one neuron to the next generates a 'propagating pulse'; thirdly, a coupled lead oscillator; or fourthly, multiple stimuli. His team has likened these phenomena to 'swarm intelligence' or 'stadium waves' emanating collectively from millions to billions of neurons. Either they can occur spontaneously, or they can be evoked as a 'stone into the puddle', or, most likely, can combine to give rich and highly variable one-off brain states from one moment to the next. Wu, J.-Y., Xiaoying Huang & Chuan Zhang. 'Propagating waves of activity in the neocortex: what they are, what they do'. *Neuroscientist*, 14, 487–502 (2008).

33. Greenfield, S. A. & Collins, T. F. T. 'A neuroscientific approach to consciousness'. *Progress in Brain Research*, 150, 11–23 (2005).

34. Collins, T. F. T. et al. 'Dynamics of neuronal assemblies are modulated by anaesthetics but not analgesics'. *European Journal of Anaesthesiology*, 24, 609–14 (2007).

35. Blumenfeld, H. 'Consciousness and epilepsy: why are patients with absence seizures absent?' *Progress in Brain Research*, 150, 271–86 (2005); see also Penfield, W. & Jasper, H. *Epilepsy and the Functional Anatomy of the Human Brain* (Little, Brown & Co., 1954).

36. Davies, D. L. & Alkana, R. L. 'Benzodiazepine agonist and inverse agonist coupling in GABAA receptors antagonized by increased atmospheric pressure'. *European Journal of Pharmacology*, 469, 37–45 (2003); see also Johnson, F. H. & Flagler, E. A. 'Hydrostatic pressure reversal of narcosis in tadpoles'. *Science*, 112, 91–2 (1950); and Johnson, F. H. & Flagler, E. A. 'Activity of narcotized amphibian larvae under hydrostatic pressure'. *Journal of Cellular Physiology*, 37, 15–25 (1951).

37. Wlodarczyk, A., McMillan, P. F. & Greenfield, S. A. 'High pressure effects in anaesthesia and narcosis'. *Chemical Society Reviews*, 35, 890–8 (2006).

38. There might also be effects of high-pressure in reversing anaesthesia at the single-cell level. Some of the original pressure reversal work was done by David White and G. H. Hulands in the late 1960s and early 1970s on single cells. White, D. C. & Halsey, M. J. 'Effects of changes in temperature and pressure during experimental anaesthesia'. *British Journal of Anaesthesia*, 46, 196–201 (1974).
39. Since there is no committed brain region for consciousness, and since conventional brain imaging shows an overall brain reduction in activity with anaesthesia, the hippocampus has so far served as a representative part of the brain for visualizing assemblies. Nonetheless, it has never really featured as part of the anatomical minimum kit when researchers have hunted for the relevant candidate brain areas. In contrast, the thalamocortical loop is just such a circuit, even though precisely how or why it plays a special part in consciousness hasn't yet been satisfactorily explained. Still, it is an obvious brain circuit to explore if we are to extend our exploration beyond isolated brain regions to ones that are connected. Alkire et al., 1999.
40. Wlodarczyk, McMillan & Greenfield, 2006.
41. Devonshire et al., 2010.
42. Collins et al., 2007.
43. Wlodarczyk, McMillan & Greenfield, 2006.
44. Devonshire et al., 2010.
45. Ibid.
46. Bryan, A. et al. 'Functional electrical impedance tomography by evoked response: a new device for the study of human brain function during anaesthesia'. *Proceedings of the Anaesthetic Research Society Meeting*, 428–9 (2010).
47. Ibid.
48. Blundon, J. A. & Zakharenko, S. S. 'Dissecting the components of long-term potentiation'. *Neuroscientist*, 14, 598–608 (2008).

3. WALKING THE DOG

1. Loh, K. K. & Kanai, R. 'Higher media multi-tasking activity is associated with smaller gray-matter density in the anterior cingulate cortex'. *PLoS One*, 9, e106698 (2014).
2. Schaefer, S. et al. 'Cognitive performance is improved while walking: differences in cognitive–sensorimotor couplings between children and young adults'. *European Journal of Developmental Psychology*, 7, 371–89 (2010).

3. Berman, M. G., Jonides, J. & Kaplan, S. 'The cognitive benefits of interacting with nature'. *Psychological Science*, 19, 1207–12 (2008).
4. Ibid.
5. Atchley, R. A., Strayer, D. L. & Atchley, P. 'Creativity in the wild: improving creative reasoning through immersion in natural settings'. *PLoS One*, 7, e51474 (2012).
6. Stourton, E. *Diary of a Dog Walker: Time Spent Following a Lead.* (Doubleday, 2011).
7. Wells, M. J. & Young, J. Z. 'The effect of splitting part of the brain or removal of the median inferior frontal lobe on touch learning in octopus'. *Journal of Experimental Biology*, 50, 515–26 (1969); see also Wells, M. J. & Young, J. Z. 'The median inferior frontal lobe and touch learning in the octopus'. *Journal of Experimental Biology*, 56, 381–402 (1972).
8. Sutherland, N. S. 'Shape discrimination in rat, octopus and goldfish: a comparative study'. *Journal of Comparative Physiological Psychology*, 67, 160–76 (1969).
9. Fiorito, G., Agnisola, C., d'Addio, M., Valanzano, A. & Calamdrei, G. 'Scopolamine impairs memory recall in Octopus vulgaris'. *Neuroscience Letters*, 253, 87–90 (1998).
10. Moriyama, T. & Gunji, Y.-P. 'Autonomous learning in maze solution by octopus'. *Ethology*, 103, 499–513 (1997).
11. Fiorito, G. & Scotto, P. 'Observational learning in Octopus vulgaris'. *Science*, 256, 545–7 (1992).
12. Giuditta, A. et al. 'Nuclear counts in the brain lobes of Octopus vulgaris as a function of body size'. *Brain Research*, 25, 55–62 (1971).
13. Herculano-Houzel, S. 'The human brain in numbers: a linearly scaled-up primate brain'. *Frontiers in Human Neuroscience*, 3, 31 (2009).
14. Juorio, A. V. 'Catecholamines and 5-hydroxytryptamine in nervous tissue of cephalopods'. *Journal of Physiology*, 216, 213–26 (1971).
15. Diamond, M. C., Krech, D. & Rosenzweig, M. R. 'The effects of an enriched environment on the histology of the rat cerebral cortex'. *Journal of Comparative Neurology*, 123, 111–20 (1964).
16. The benefits of an enriched environment: increased cell-body size of neurons; increased overall brain weight; increased thickness of cortex; greater number of dendritic 'spines' (the tiny protuberances on branches of cells that allow for highly specific contacts); increase in size of synaptic junctions, hence of connections; and an increase in the number of glial cells – the housekeeping cell that ensures a benign micro-environment for the key players: the neurons.

17. Valero, J. et al., 'Short-term environmental enrichment rescues adult neurogenesis and memory deficits in APPSw,Ind transgenic mice'. PLoS One, 6, 2 (2011).
18. Speisman, R. B. et al. 'Environmental enrichment restores neurogenesis and rapid acquisition in aged rats'. *Neurobiology of Aging*, 34, 263–74 (2013).
19. van Dellen, A. et al. 'Delaying the onset of Huntington's in mice'. *Nature*, 404, 721–2 (2000).
20. Young, D. et al. 'Environmental enrichment inhibits spontaneous apoptosis, prevents seizures and is neuroprotective'. *Nature Medicine*, 5, 448–53 (1999); see also Johansson, B. B. 'Functional outcome in rats transferred to an enriched environment 15 days after focal brain ischemia'. *Stroke*, 27, 324–6 (1996).
21. Amaral, O. B. et al. 'Duration of environmental enrichment influences the magnitude and persistence of its behavioral effects on mice'. *Physiology & Behavior*, 93, 388–94 (2008).
22. It seems that environmental enrichment may also have more subtle effects. Mink housed in non-enriched cages spent more time interacting with novel objects, irrespective of whether they are aversive, rewarding and neutral, when they are finally introduced to their cage: the newly privileged mink made contact with these objects more quickly than did animals housed in enriched cages. The conclusion from these observations was one of an apparently greater 'boredom' in non-enriched animals. We do need to be careful not to over-interpret these results – hence the scare quotes: whenever we use terms most appropriate for the human behavioural repertoire, it might be prudent not to take the term too literally. Mink are not humans, and we cannot ascribe to them a sophisticated mental state such as boredom without qualification. Suffice it to say that the enriched environment was such a completely novel experience in itself that it dwarfed the 'novelty' of individual components and offered a much greater differential when compared to their previous lifestyle than for animals used to being presented with novel objects less routinely. However, a taste of enrichment may be worse than none at all. By not being able to engage in motivated behaviours, some caged animals exhibit stereotypy: repetitive, purposeless movements often indicating high stress. In such animals, enrichment can reduce these movements, but, after being housed in an enriched environment, animals can exhibit worse stereotypy than those housed continuously in non-enriched cages. Meagher, R. K. & Mason, G. J. 'Environmental enrichment reduces signs of boredom in

caged mink'. *PLoS One*, 7, e49180 (2012). See also Latham, N. & Mason, G. 'Frustration and perseveration in stereotypic captive animals: is a taste of enrichment worse than none at all?' *Behavioural Brain Research*, 211, 96–104 (2010).

23. Mora, F., Segovia, G. & del Arco, A. 'Aging, plasticity and environmental enrichment: structural changes and neurotransmitter dynamics in several areas of the brain'. *Brain Research Review*, 55, 78–88 (2007). Kozorovitskiy, Y. et al. 'Experience induces structural and biochemical changes in the adult primate brain'. *Proceedings of the National Academy of Sciences of the United States of America*, 102, 17478–82 (2005).
24. Kolb, B. *Brain, Plasticity and Behaviour*, ch 1. Laurence Erlbaum Assoc (1995)
25. Greenfield, S. A. *Mind Change: How Digital Technologies are Leaving Their Mark on Our Brains*. (Random House, 2014).
26. In addition, there is an increase in synapse density between twenty-eight weeks' gestation and seventy weeks', namely, some thirty weeks after birth. Synapse density reaches a peak with 600 million synapses in a cubic millimetre at eight months, before stabilizing at about ten years of age to 300 million. Huttenlocher, P. et al. 'Synaptogenesis in human visual cortex – evidence for synapse elimination during normal development'. *Neuroscience Letters*, 33, 247–52 (1982).
27. Gogtay, N. et al. 'Dynamic mapping of human cortical development during childhood through early adulthood'. *Proceedings of the National Academy of Sciences of the United States of America*, 101, 8174–9 (2004).
28. Meanwhile, cells in 'association' areas (those areas of cortex not primarily involved in sensory or motor processing), such as the prefrontal cortex, undergo a more prolonged decline than, say, the sensory cortex, where the sharpness of the growth and decline phases are thought to underlie 'critical periods' in sensory modalities, when, as the term suggests, time frames are crucial in establishing the correct circuitry. Interestingly, the volume of 'white matter' – that is, the connecting fibres – continues to increase over the entire time period investigated in this study, due to increased myelination – the insulation that improves nerve conductance and accounts for the impairments seen in multiple sclerosis, a devastating disease, in which it deteriorates.
29. Maguire, E. A. et al. 'Navigation-related structural change in the hippocampi of taxi drivers'. *Proceedings of the National Academy of Sciences of the United States of America*, 97, 4398–403 (2000).
30. Gaser, C. & Schlaug, G. 'Brain structures differ between musicians and non-musicians'. *Journal of Neuroscience*, 23, 9240–5 (2003).

31. Bengtsson, S. L. et al. 'Extensive piano practising has regionally specific effects on white matter development'. *Nature Neuroscience*, 8, 1148–50 (2005).
32. Jäncke, L. et al. 'The architecture of the golfer's brain'. *PLoS One*, 4, e4785 (2009).
33. Park, I. S. et al. 'Experience-dependent plasticity of cerebellar vermis in basketball players'. *Cerebellum*, 8, 334–9 (2009).
34. Mechelli, A. et al. 'Neurolinguistics: structural plasticity in the bilingual brain'. *Nature*, 431, 757 (2004); see also Stein, M. et al. 'Structural plasticity in the language system related to increased second-language proficiency'. *Cortex*, 48, 458–65 (2012).
35. Draganski, B. et al. 'Neuroplasticity: changes in grey matter induced by training'. *Nature*, 427, 311–12 (2004); see also Driemeyer, J. et al. 'Changes in gray matter induced by learning – revisited'. *PLoS One*, 3, e2669 (2008).
36. Pascual-Leone, A. et al. 'Modulation of muscle responses evoked by transcranial magnetic stimulation during the acquisition of new fine motor skills'. *Journal of Neurophysiology*, 74, 1037–45 (1995).
37. Bailey, C. H. & Kandel, E. R. 'Synaptic remodeling, synaptic growth and the storage of long-term memory in Aplysia'. *Progress in Brain Research*, 169, 179–98 (2008).
38. Ridley, M. *Nature via Nurture: Genes, Experience and What Makes Us Human*. (Harper Perennial, 2004).
39. Greenfield, 2014; see also Greenfield, S. A. *You and Me: The Neuroscience of Identity*. (Notting Hill Editions, 2011).
40. Pittenger, C. & Kandel, E. R. 'In search of general mechanisms for long-lasting plasticity: Aplysia and the hippocampus'. *Philosophical Transactions of the Royal Society of London B: Biological Sciences*, 358, 757–63 (2003); for a basic explanation of Long Term Potentiation (LTP) and Long Term Depresssion (LTD) see S. A. Greenfield, *The Human Brain: A Guided Tour* (Orion, 1997); for a fuller and more technical description see D. Purves, *Neuroscience* (Sinauer Press, 2011, 5th edn).
41. Deidda, G., Bozarth, I. F. & Cancedda, L. 'Modulation of GABAergic transmission in development and neurodevelopmental disorders: investigating physiology and pathology to gain therapeutic perspectives'. *Frontiers in Cellular Neuroscience*, 8, 119 (2014).
42. Storer, K. P. & Reeke, G. N. 'γ-Aminobutyric acid receptor type A receptor potentiation reduces firing of neuronal assemblies in a computational cortical model'. *Anesthesiology*, 117, 780–90 (2012).

43. Olds, J. & Milner, P. 'Positive reinforcement produced by electrical stimulation of septal area and other regions of rat brain'. *Journal of Comparative Physiological Psychology*, 47, 419–27 (1954).
44. Dopamine is inhibitory on the prefrontal cortex. See Ferron, A. et al. 'Inhibitory influence of the mesocortical dopaminergic system on spontaneous activity or excitatory response induced from the thalamic mediodorsal nucleus in the rat medial prefrontal cortex'. *Brain Research*, 302, 257–65 (1984). Gao, W.-J., Wang, Y. & Goldman-Rakic, P. S. 'Dopamine modulation of perisomatic and peridendritic inhibition in prefrontal cortex'. *Journal of Neuroscience*, 23, 1622–30 (2003). Cole, M. W. & Schneider, W. 'The cognitive control network: integrated cortical regions with dissociable functions'. *Neuroimage*, 37, 343–60 (2007). Cole, M. W., Pathak, S. & Schneider, W. 'Identifying the brain's most globally connected regions'. *NeuroImage*, 49, 3132–48 (2010). Cools, R. & d'Esposito, M. 'Inverted-U-shaped dopamine actions on human working memory and cognitive control'. *Biological Psychiatry*, 69, e113–25 (2011). Thanks to this inhibitory action, dopamine could restrict assembly size by working at three different levels. *Firstly*, at the level of single neurons, by a direct modulatory effect, causing a well-established inhibition of the neurons in a particular brain region, the prefrontal cortex. *Secondly*, by indirect anatomical systems in the holistic brain: the prefrontal cortex connects to more areas of the brain than any other. Hence, if dopamine inhibits such a pivotal area, particularly in humans, then the overall organization of the brain will become more fragmented, and hence any assembly that transcends normal anatomical compartmentalization will be less likely to form. *Thirdly*, and finally, dopamine could exert an effect on reducing assembly size at an indirect, behavioural level. Drugs such as amphetamine, which release dopamine in the brain, are well known to have overall stimulant effects, resulting in hyperactivity. In such a state, when dopamine levels are high, there will be an increased likelihood that the hyperactive user will have more opportunity for immediate sensation from the environment around them, rather than have a prolonged, pensive 'cognitive' take on one object or person: a new stimulation is sought, a new stone is effectively thrown into the puddle, before the excursions from its predecessor have a chance to generate the full potential radius of the original ripples: this will be a world, as we have seen, of rapidly competing and therefore reduced-size assemblies, due to a high turnover.
45. See also a comparable effect with dopamine in Susanta Bandyopadhyay and John J. Hablitz (2007), 'Dopaminergic Modulation of Local

Network Activity in Rat Prefrontal Cortex', *Journal of Neurophysiology*: 4120–28.

46. Brown, R. T. & Wagner, A. R. 'Resistance to punishment and extinction following training with shock or nonreinforcement'. *Journal of Experimental Psychology*, 68, 503–7 (1964); see also Gray, J. A. 'Fear, panic and anxiety: what's in a name. *Psychological Inquiry*, 2, 72–96 (1991).
47. Greenfield, S. A. *The Private Life of the Brain: Emotions, Consciousness and the Secret of the Self.* (Wiley, 2000).
48. http://www.channel4.com/news/laughing-gas-nitrous-oxide-legal-high-police-drugs-brick-lane-festivals
49. Opiates tend to be inhibitory but often act on inhibitory circuits to produce disinhibition. However, such an excitatory effect could, similarly, produce a small-assembly state through desynchronization. Charles, A. C. & Hales, T. G. 'From inhibition to excitation: functional effects of interaction between opioid receptors'. *Life Sciences*, 76, 479–85 (2004).
50. Yau, S. et al. 'Physical exercise-induced adult neurogenesis: a good strategy to prevent cognitive decline in neurodegenerative diseases?' *Biomedical Research International*, 2014, 403120 (2014).
51. Olson, A. K. et al 'Environmental enrichment and voluntary exercise massively increase neurogenesis in the adult hippocampus via dissociable pathways'. *Hippocampus*, 16, 250–60 (2006).
52. van Praag, H., Kempermann, G. & Gage, F. H. 'Running increases cell proliferation and neurogenesis in the adult mouse dentate gyrus'. *Nature Neuroscience*, 2, 266–70 (1999).
53. Begley, S. *The Plastic Mind.* (Constable, 2009).
54. He, S.-B. et al. 'Exercise intervention may prevent depression'. *International Journal of Sports Medicine*, 33, 525–30 (2012).
55. Frodl, T. & O'Keane, V. 'How does the brain deal with cumulative stress? A review with focus on developmental stress, HPA axis function and hippocampal structure in humans'. *Neurobiology of Disease*, 52, 24–37 (2013).
56. Cakır, B. et al. 'Stress-induced multiple organ damage in rats is ameliorated by the antioxidant and anxiolytic effects of regular exercise'. *Cell Biochemistry and Function*, 28, 469–79 (2010). Schoenfeld et al., 2013, found both new excitatory and inhibitory cells in the hippocampus in running animals; also, the running animals would experience shorter stressful episodes after a stressor: 'the hippocampus of runners is vastly different from that of sedentary animals. Not only are there more excitatory neurons and more excitatory synapses, but the inhibitory neurons are more likely to become activated, presumably to dampen the excitatory neurons, in response to stress.' Schoen-

feld T. J. et al. 'Physical exercise prevents stress-induced activation of granule neurons and enhances local inhibitory mechanisms in the dentate gyrus'. *Journal of Neuroscience*, 33(18): 7770–7 (2013). Another interesting study has shown that moderate exercise versus quiet rest can help buffer against future stress (Smith, 2013). Though both reduced stress in the short term, the subjects in the rest condition were more stressed by emotional pictures than the exercising group. Smith, J. C. 'Effects of emotional exposure on state anxiety after acute exercise'. *Medicine and Science in Sports and Exercise*, 45(2): 372–8 (2013); and Nakajima, S. et al. 'Regular voluntary exercise cures stress-induced impairment of cognitive function and cell proliferation accompanied by increases in cerebral IGF-1 and GST activity in mice'. *Behavioural Brain Research*, 211, 178–84 (2010).

57. Nokia, M. et al. 'Learning to learn: theta oscillations predict new learning, which enhances related learning and neurogenesis'. *PLoS One*, 7, e31375 (2012).
58. Begley, 2009.
59. Devonshire, I. M. et al. 'Environmental enrichment differentially modifies specific components of sensory-evoked activity in rat barrel cortex as revealed by simultaneous electrophysiological recordings and optical imaging in vivo'. *Neuroscience*, 170, 662–9 (2010).
60. Ibid.
61. Assemblies in the awake mouse compared to the anaesthetized counterpart are larger in terms of amplitude and (much larger) in terms of spatial extent. Ferezou I, Bolea S. & Petersen C. C. 'Visualizing the cortical representation of whisker touch: voltage-sensitive dye imaging in freely moving mice'. *Neuron*, 50, 617–29 (2006).
62. Greenfield, 2011.

4. BREAKFAST

1. Reich, L. et al. 'A ventral visual stream reading center independent of visual experience'. *Current Biology*, 21, 363–8 (2011).
2. Neville, H. J. & Lawson, D. 'Attention to central and peripheral visual space in a movement detection task: an event-related potential and behavioral study. II. Congenitally deaf adults'. *Brain Research*, 405, 268–83 (1987).
3. Karns, C. M., Dow, M. W. & Neville, H. J. 'Altered cross-modal processing in the primary auditory cortex of congenitally deaf adults: a

visual-somatosensory fMRI study with a double-flash illusion'. *Journal of Neuroscience*, 32, 9626–38 (2012).

4. Gougoux, F. et al. 'Neuropsychology: pitch discrimination in the early blind'. *Nature*, 430, 309 (2004).
5. Lessard, N. et al. 'Early-blind human subjects localize sound sources better than sighted subjects'. *Nature*, 395, 278–80 (1998).
6. Röder, B. et al. 'Semantic and morpho-syntactic priming in auditory word recognition in congenitally blind adults'. *Language and Cognitive Processes*, 18, 1–20 (2003).
7. Bull, R., Rathborn, H. & Clifford, B. R. 'The voice-recognition accuracy of blind listeners'. *Perception*, 12, 223–6 (1983).
8. Petrus, E. et al. 'Crossmodal induction of thalamocortical potentiation leads to enhanced information processing in the auditory cortex'. *Neuron*, 81, 664–73 (2014).
9. Terhune, D. B. et al. 'Enhanced cortical excitability in grapheme-color synesthesia and its modulation'. *Current Biology*, 21, 2006–9 (2011).
10. Liotta, A. et al. 'Partial disinhibition is required for transition of stimulus-induced sharp wave-ripple complexes into recurrent epileptiform discharges in rat hippocampal slices'. *Journal of Neurophysiology*, 105, 172–87 (2011).
11. Rockel, A. J., Hiorns, R. W. & Powell, T. P. 'The basic uniformity in structure of the neocortex'. *Brain*, 103, 221–44 (1980). On the other hand, some might argue that the cookie-cutter model of the cortex cytoarchitecture doesn't always hold, and it could all be explained by various anatomical differences in visual and auditory pathways. It could well be that the differences observed in the imaging data described simply reflects a different distribution pattern in inputs to the auditory, compared to the visual, cortex. For example, the auditory cortex usually receives two-ear input from the brain areas below the cortex, whereas the visual cortex receives inputs from only two visual hemi-fields. Furthermore, the main inputs from the subcortical relay station (the thalamus) to the visual cortex terminate primarily in one layer of cortex (layer IV), whereas in the auditory cortex some thalamic inputs are distributed more widely (extending well into layer III), while others terminate elsewhere (in layer I). LeVay, S. & Gilbert, C. D. 'Laminar patterns of geniculocortical projection in the cat'. *Brain Research*, 113(1), 1–19 (1976). Smith, P. H. & Populin, L. C. 'Fundamental differences between the thalamocortical recipient layers of the cat auditory and visual cortices'. *Journal of Comparative Neurology*, 436(4), 508–19 (2001). Huang, C. L. & Winer, J. A. 'Auditory thalam-

ocortical projections in the cat: laminar and areal patterns of input'. *Journal of Comparative Neurology*, 427(2), 302–31 (2000).

12. Libet, B. *Mind Time: The Temporal Factor in Consciousness*. (Harvard University Press, 2004).
13. VanRullen, R. & Thorpe, S. J. 'The time course of visual processing: from early perception to decision-making'. *Journal of Cognitive Neuroscience*, 13, 454–61 (2001).
14. Chakraborty, S., Sandberg, A. & Greenfield, S. A. 'Differential dynamics of transient neuronal assemblies in visual compared to auditory cortex'. *Experimental Brain Research*, 182, 491–8 (2007).
15. Libet, 2004.
16. Crick, F. & Koch, C. 'A framework for consciousness'. *Nature Neuroscience*, 6, 119–26 (2003).
17. Dunn, R. & Dunn, K. *Teaching Students through Their Individual Learning Styles: A Practical Approach*. (Prentice Hall, 1978).
18. Pashler, H. et al. 'Learning Styles: Concepts and Evidence'. *Psychological Science in the Public Interest*, 9, 105–119 (2009).
19. Dehaene, S. et al. 'Arithmetic and the brain'. *Current Opinions in Neurobiology*, 14, 218–24 (2004).
20. Grady, D. 'The vision thing: mainly in the brain'. *Discover* (http://discovermagazine.com/1993/jun/thevisionthingma227). Similarly, see the website of Nikos Logothetis: http://www.kyb.tuebingen.mpg.de/research/dep/lo/visual-perception.html.
21. Jones, B. 'Spatial perception in the blind'. *British Journal of Psychology*, 66, 461–72 (1975).
22. Calvert, G. A., Campbell, R. & Brammer, M. J. 'Evidence from functional magnetic resonance imaging of crossmodal binding in the human heteromodal cortex'. *Current Biology*, 10, 649–57 (2000).
23. Risberg, A. & Lubker, J. 'Prosody and speech-reading' in *STL Quarterly Progress and Status Report*, 4, 1–16 (1978). This experimental sensory synergy has been shown more directly, that is, invasively, in monkeys, where the process of sensory integration need not be attributed to classical 'association cortices' (the 'higher' areas not directly linked to specific sensory modalities): instead, early-stage multisensory influences can enhance the information carried by neurons. While the five senses remain identifiable as such, they will interact and, to complicate things further, have different possible net effects according to how the relevant neurons happen to be wired to each other. Such intricate wiring patterns would not be detectable from simple brain imaging but would require the more finely tuned and detailed technique of electrophysiology.

24. Kayser, C., Petkov, C. I. & Logothetis, N. K. 'Multisensory interactions in primate auditory cortex: fMRI and electrophysiology'. *Hearing Research*, 258, 80–8 (2009).
25. Geake, J. Ch. 2, pp. 10–17 in *Companion to Gifted Education* (eds. Balchin, T. & Hymer, B.) (Routledge, 2008).
26. Bakalar, N. 'Sensory science: partners in flavour'. *Nature*, 486, S4–5 (2012).
27. Small, D. M. 'How does food's appearance or smell influence the way it tastes?' *Scientific American*, 299, 100 (2008).
28. Stevenson, R. J., Prescott, J. & Boakes, R. A. 'Confusing tastes and smells: how odours can influence the perception of sweet and sour tastes'. *Chemical Senses*, 24, 627–35 (1999).
29. Spence, C. & Piqueras-Fiszman, B. *The Perfect Meal: The Multisensory Science of Food and Dining.* (Wiley, 2014).
30. Gal, D., Wheeler, S. C. & Shiv, B. 'Cross-modal influences on gustatory perception'. (2007).
31. Harrar, V. & Spence, C. 'The taste of cutlery: how the taste of food is affected by the weight, size, shape and colour of the cutlery used to eat it'. *Flavour*, 2, 21 (2013).
32. Spence, C. 'Auditory contributions to flavour perception and feeding behaviour'. *Physiology and Behavior*, 107, 505–15 (2012).
33. Chen, J. & Eaton, L. 'Multimodal mechanisms of food creaminess sensation'. *Food and Function*, 3, 1265–70 (2012); see also Green, B. G. & Nachtigal, D. 'Somatosensory factors in taste perception: effects of active tasting and solution temperature'. *Physiology and Behavior*, 107, 488–95 (2012).
34. Harrar & Spence, 2013.
35. Cruz, A. & Green, B. G. 'Thermal stimulation of taste'. *Nature*, 403, 889–92 (2000).
36. Lindstrom, M. *Brand Sense: Sensory Secrets behind the Stuff We Buy.* (Free Press, 2010).
37. Geertz, C. *The Interpretation of Cultures: Selected Essays.* (Basic Books, 1973).
38. Lawless, H. T. & Heymann, H. *Sensory Evaluation of Food: Principles and Practices.* (Springer, 2010).
39. Lindstrom, 2010.
40. Lubke, G. H. et al. 'Dependence of explicit and implicit memory on hypnotic state in trauma patients'. *Anesthesiology*, 90, 670–80 (1999); see also Kerssens, C. et al. 'Auditory information processing during adequate propofol anesthesia monitored by electroencephalogram bispectral index'. *Anesthesia and Analgesia*, 92, 1210–14 (2001).

41. Nordin, S. et al. 'Evaluation of auditory, visual and olfactory event-related potentials for comparing interspersed- and single-stimulus paradigms'. *International Journal of Psychophysiology*, 81, 252–62 (2011).
42. Sun, G. H. et al. 'Olfactory identification testing as a predictor of the development of Alzheimer's dementia: a systematic review'. *Laryngoscope*, 122, 1455–62 (2012).
43. Wilson, D. A., Kadohisa, M. & Fletcher, M. L. 'Cortical contributions to olfaction: plasticity and perception'. *Seminars in Cell & Developmental Biology*, 17, 462–70 (2006).
44. Wysocki, C. J. & Preti, G. 'Facts, fallacies, fears and frustrations with human pheromones'. *The Anatomical Record. Part A. Discoveries in Molecular, Cellular and Evolutionary Biology*, 281, 1201–11 (2004).
45. Porter, R. H., Cernoch, J. M. & Balogh, R. D. 'Odor signatures and kin recognition'. *Physiology and Behavior*, 34, 445–8 (1985).
46. Weisfeld, G. E. et al. 'Possible olfaction-based mechanisms in human kin recognition and inbreeding avoidance'. *Journal of Experimental Child Psychology*, 85, 279–95 (2003).
47. Wedekind, C. 'Body odours and body odour preferences in humans' in *The Oxford Handbook of Evolutionary Psychology* (eds. Dunbar, R. & Barrett, L.) (Oxford University Press, 2007).
48. Lewis, P. 'Musical minds'. *Trends in Cognitive Science*, 6, 364 (2002).
49. *Mousterian 'Bone Flute' and Other Finds from Divje Babe I Cave Site in Slovenia*. (Znanstvenoraziskovalni Centre Sazu, 2007).
50. Huron, D. 'Is music an evolutionary adaptation?' *Annals of the New York Academy of Sciences*, 930, 43–61 (2001).
51. Barrow, J. D. *The Artful Universe. The Biological Foundations of Music* (Clarendon Press, 1995).
52. Pinker, S. *How the Mind Works*. (W. W. Norton, 1997).
53. Ibid.
54. Molino, J. 'Toward an evolutionary theory of music and language' in *The Origins of Music* (eds. Wallin, N., Merker, B. & Brown, S.), 165–76 (MIT Press, 2000).
55. Huron, 2001. A common perception is that, in Islamic tradition, music could be a distraction from the important things of life and, worse still, lead to licentious behaviour. However, as John Baily, a Radio Liberty correspondent, has pointed out, 'It isn't actually correct to say that the Taliban have banned music. They have banned musical instruments and any kind of music-making that involves musical instruments, quite possibly with one exception, and that exception is the frame drum –

the duff – because there are hadiths in which the Prophet Muhammad appreciates or allows the frame drum to be used in connection with celebrations of weddings and so on.' Baily, J., *British Journal of Ethnomusicology*: 'It isn't actually correct to say Taliban have banned music'. Radio Liberty, 22 June 2009.

56. Merriam, A. P. *The Anthropology of Music*. (Northwestern University Press, 1964).
57. Cross, I. 'Music, cognition, culture and evolution'. *Annals of the New York Academy of Sciences*, 930, 28–42 (2001); see also Storr, A. *Music and the Mind*. (Random House, 1992).
58. Dunbar, R. *Human Evolution. Music, Cognition, Culture and Evolution* (Pelican, 2014).
59. McNeill, W. H. *Keeping Together in Time*. (Harvard University Press, 1995).
60. Cross, 2001.
61. Trevarthen, C. 'Musicality and the intrinsic motive pulse: evidence from human psychobiology and infant communication'. *Musicae Scientiae*, 155–215 (2000).
62. Cross, 2001.
63. The amygdala is so named from the Latin for its almond-like shape and is situated deep in the temporal lobes, forming part of the limbic system; it has connections to diverse brain regions. As such, it has been implicated in a variety of functions related to both emotion and memory. For two recent reviews see LaLumiere, R. T. 'Optogenetic dissection of amygdala functioning'. *Frontiers in Behavioral Neuroscience*, 8, 107 (2014) and Fernando, A. B. P., Murray, J. E. & Milton, A. L. 'The amygdala: securing pleasure and avoiding pain'. *Frontiers in Behavioral Neuroscience*, 7, 190 (2013). Also, of specific relevance to music, see Griffiths, T. D. et al. '"When the feeling's gone": a selective loss of musical emotion'. *Journal of Neurology, Neurosurgery & Psychiatry*, 75, 344–5 (2004).
64. Gosselin, N. et al. 'Impaired recognition of scary music following unilateral temporal lobe excision'. *Brain*, 128, 628–40 (2005).
65. Panksepp, J. 'The emotional sources of "chills" induced by music'. *Music Perception*, 13, 171–207 (1995).
66. Salimpoor, V. N. et al. 'Anatomically distinct dopamine release during anticipation and experience of peak emotion to music'. *Nature Neuroscience*, 14, 257–62 (2011).
67. Wise, R. A. 'Forebrain substrates of reward and motivation'. *Journal of Comparative Neurology*, 493, 115–21 (2005).

68. Burton, A. C., Nakamura, K. & Roesch, M. R. 'From ventral-medial to dorsal-lateral striatum: neural correlates of reward-guided decision-making'. *Neurobiology of Learning and Memory*, 117, 51–9 (2014).
69. Small, D. M. et al. 'Changes in brain activity related to eating chocolate: from pleasure to aversion'. *Brain*, 124, 1720–33 (2001).
70. Breiter, H. C. et al. 'Acute effects of cocaine on human brain activity and emotion'. *Neuron*, 19, 591–611 (1997).
71. Zald, D. H. & Zatorre, Robert J., C. 19, 'Music', pp. 405–28 in *Neurobiology of Sensation and Reward* (ed. Gottfried, J. A.) (CRC Press, 2011).
72. Benoit, C.-É. et al. 'Musically cued gait-training improves both perceptual and motor timing in Parkinson's disease'. *Frontiers in Human Neuroscience*, 8, 494 (2014); see also Nombela, C. et al. 'Into the groove: can rhythm influence Parkinson's disease?' *Neuroscience and Biobehavioral Reviews*, 37, 2564–70 (2013); and Pacchetti, C. et al. 'Active music therapy in Parkinson's disease: an integrative method for motor and emotional rehabilitation'. *Psychosomatic Medicine*, 62, 386–93 (2000).
73. Azulay, J. P. et al. 'Visual control of locomotion in Parkinson's disease. *Brain*, 122 (Pt 1, 111–20 (1999).
74. Lim, I. et al. 'Effects of external rhythmical cueing on gait in patients with Parkinson's disease: a systematic review'. *Clinical Rehabilitation*, 19, 695–713 (2005). However, despite evidence suggesting that internal cues are indeed preferentially effected by the disease over external cues, successful interventions perfected under laboratory conditions don't generally provide improvements outside the laboratory, that is, in patients' homes.
75. In any event, once dopamine had been identified as surging during musical experiences, Salimpoor and her team used a much faster but less chemical-specific scanning technique, Positron Emission Tomography (PET), to monitor the course over time of the dopamine release from key areas. They found a functional dissociation: one brain region (the caudate) was more involved during anticipation of the music's power, while another (the nucleus accumbens) was more related to the experience of peak emotional responses to music, with the two events being separated by some ten to fifteen seconds.
76. Meyer, L. *Emotion and Meaning in Music*. (University of Chicago Press, 1956).
77. Blood, A. J. & Zatorre, R. J. 'Intensely pleasurable responses to music correlate with activity in brain regions implicated in reward and emotion'. *Proceedings of the National Academy of Sciences of the United States of America*, 98, 11818–23 (2001).

78. Sacks, O. *Musicophilia: Tales of Music and the Brain.* (Random House, 2007).

5. AT THE OFFICE

1. http://www.bls.gov/tus/charts/
2. One large-scale review of the possible therapeutic effects of certain types of hospital design over others included 102 studies and 38 meta-analyses. Yet the lead author, Dr Amy Drahota, from the University of Portsmouth, concluded: 'Although it may seem obvious that the area in which a patient recovers will affect his or her healing, we need research evidence to help guide all the decisions in hospital design. However, 85 of our included studies related to the use of music in hospital. We were also interested in finding studies on all aspects of hospital environment design, but many areas have not had high-quality research studies. For example, we only found one study that met our criteria for lighting, one study that met our criteria for decoration, and we didn't find any studies of high-enough quality which looked at the use of art, or way-finding aids. And we even have some reservations about the studies we did find because their quality is not as good as it could be. So, despite the large volume of information included in this review, we're calling for more high-quality studies looking at the different components of hospital design in order to make better-informed decisions on how to design or refurbish our hospitals in the future.' http://www.news-medical.net/news/20120315/Hospital-environments-could-influence-patient-recovery.aspx. Drahota, A. et al. 'Sensory environment on health-related outcomes of hospital patients'. Cochrane Database of Systematic Reviews, 3, CD005315 (2012). See also Kaler, S. R. & Freeman, B. J. 'Analysis of environmental deprivation: cognitive and social development in Romanian orphans'. *Journal of Child Psychology and Psychiatry*, 35, 769–81 (1994). Eluvathingal, T. J. et al. 'Abnormal brain connectivity in children after early severe socioemotional deprivation: a diffusion tensor imaging study'. *Pediatrics*, 117(6), 2093–100 (June 2006). Prut, L. & Belzung, C. 'The open field as a paradigm to measure the effects of drugs on anxiety-like behaviors: a review'. *European Journal of Pharmacology*, 463(1–3), 3–33 (28 Feb. 2003); see also Eberhard, J. P. 'Applying neuroscience to architecture'. *Neuron*, 62, 753–6 (2009). Eberhard makes an interesting point in the article cited about the existence of defined and specific hypotheses on human responses to architecture, which, nonetheless, are simply waiting to be researched by students and

research staff. In the article he claims that there are seventy to eighty such hypotheses (and he expands on them in his book *Brain Landscapes* Oxford University Press, 2009), such as the brain being hard-wired to respond to proportions based on the golden mean and a network in the brain being responsible for a sense of awe. In an attempt to address the shortcomings of previous studies and the functional requirements of a range of buildings such as hospitals, schools, sacred places, laboratories, and so on, the Academy of Neuroscience for Architects (ANFA) was created in 2003: it stands testimony to the enormous potential, as yet not completely realized, of applying neuroscience to architecture.

3. Johnson, D. E. et al. 'Growth and associations between auxology, caregiving environment and cognition in socially deprived Romanian children randomized to foster vs ongoing institutional care'. *Archives of Pediatrics and Adolescent Medicine*, 164, 507–16 (2010).
4. Albers, L. H. et al. 'Health of children adopted from the former Soviet Union and Eastern Europe. Comparison with preadoptive medical records'. *JAMA*, 278, 922–4 (1997).
5. Kaler & Freeman, 1994.
6. Eluvathingal, 2006.
7. Sheridan, M. A. et al. 'Variation in neural development as a result of exposure to institutionalization early in childhood'. *Proceedings of the National Academy of Sciences of the United States of America*, 109, 12927–32 (2012); see also Nelson, C. A. et al. 'Cognitive recovery in socially deprived young children: the Bucharest Early Intervention Project'. *Science*, 318, 1937–40 (2007).
8. Frasca, D. et al. 'Traumatic brain injury and post-acute decline: what role does environmental enrichment play? A scoping review'. *Frontiers of Human Neuroscience*, 7, 31 (2013).
9. Linhares, J. M. M., Pinto, P. D. & Nascimento, S. M. C. 'The number of discernible colors in natural scenes'. *Journal of the Optical Society of America A. Optics, Image Science and Vision*, 25, 2918–24 (2008).
10. Gegenfurtner, K. R. & Kiper, D. C. 'Color vision'. *Annual Review of Neuroscience*, 26, 181–206 (2003).
11. Chalmers, D. 'Absent qualia, fading qualia, dancing qualia' in *Conscious Experience* (ed. Metzinger, T.) (Imprint Academic, 1995).
12. Purves, D. et al. *Neuroscience.* (Sinauer Press, 2012).
13. This phenomenon is known as 'chromostereopsis': where a red patch of colour, for example, appears nearer than a blue one. Arguably, a red object, if nearer, will demand more attention and therefore have the chance of a directly physical effect that is stimulating and raises the pulse

rate. In contrast, a blue object will not appear to be as close to us as an adjacent red one and will therefore be relatively soothing, thereby aiding concentration. Cauquil, A. S. et al. 'Neural correlates of chromostereopsis: an evoked potential study'. *Neuropsychologia*, 47, 2677–81 (2009). Dreiskaemper, D. et al. 'Influence of red jersey color on physical parameters in combat sports'. *Journal of Sport and Exercise Psychology*, 35, 44–9 (2013). Farrelly, D. et al. 'Competitors who choose to be red have higher testosterone levels'. *Psychological Science*, 24, 2122–4 (2013).

14. Vandewalle, G. et al. 'Spectral quality of light modulates emotional brain responses in humans'. *Proceedings of the National Academy of Sciences of the United States of America*, 107, 19549–54 (2010).
15. Sable, P. & Akcay, O. 'Response to colour: literature review with cross-cultural marketing perspective. *International Bulletin of Business Administration*, 11, 34–41 (2011).
16. Labrecque, L. I. & Milne, G. R. 'Exciting red and competent blue: the importance of color in marketing'. *Journal of the Academy of Marketing Science*, 40, 711–27 (2012); see also Labrecque, L. I. & Milne, G. R. 'To be or not to be different: exploration of norms and benefits of color differentiation in the marketplace'. *Marketing Letters*, 24, 165–76 (2013).
17. Bottomley, P. A. 'The interactive effects of colors and products on perceptions of brand logo appropriateness'. *Marketing Theory*, 6, 63–83 (2006); see also Hanss, D., Böhm, G. & Pfister, H. R. 'Active red sports car and relaxed purple-blue van: affective qualities predict color appropriateness for car types'. *Journal of Consumer Behaviour*, 11, 368–80 (2012); and Ngo, M. K., Piqueras-Fiszman, B. & Spence, C. 'On the colour and shape of still and sparkling water: insights from online and laboratory-based testing'. *Food Quality and Preference*, 24, 260–8 (2012).
18. For an extended review and discussion of the 'red effect' see Elliot, A. J. & Maier, M. A. 'Color psychology: effects of perceiving color on psychological functioning in humans'. *Annual Review of Psychology*, 65, 95–120 (2014).
19. Mehta, R. & Zhu, R. J. 'Blue or red? Exploring the effect of color on cognitive task performances'. *Science*, 323, 1226–9 (2009); see also Elliot & Maier, 2014.
20. Mehta & Zhu, 2009.
21. Ibid.
22. Lichtenfeld, S. et al. 'Fertile green: green facilitates creative performance'. *Personality and Social Psychology Bulletin*, 38, 784–97 (2012).
23. Wallach, M. A. & Kogan, N. 'A new look at the creativity–intelligence distinction'. *Journal of Personality*, 33, 348–69 (1965).

24. http://www.morganlovell.co.uk/articles/the-evolution-of-office-design
25. Meyerson, J., & Ross, P., *The Twenty-first Century Office* (Laurence King, 2003).
26. Thanem, T., Varlander, S. & Cummings, S. 'Open space = open minds? The ambiguities of pro-creative office design'. *International Journal of Work Organisation and Emotion*, 4, 78 (2011).
27. Ibid.
28. http://www.economist.com/blogs/schumpeter/2014/05/hot-desking-and-office-hire
29. Prut, L. & Belzung, C. 'The open field as a paradigm to measure the effects of drugs on anxiety-like behaviors: a review'. *European Journal of Pharmacology*, 463, 3–33 (2003).
30. Mayo, E. *Hawthorne and the Western Electric Company: The Social Problems of an Industrial Civilization*. (Routledge, 1949).
31. Toker, U. & Gray, D. O. 'Innovation spaces: workspace planning and innovation in US university research centers'. *Research Policy*, 37, 309–29 (2008).
32. Backhouse, A. & Drew, P. 'The design implications of social interaction in a workplace setting'. *Environment and Planning B*, 19, 573–84 (1992).
33. Toker & Gray, 2008.
34. Greenfield, S. A. *You and Me: The Neuroscience of Identity*. (Notting Hill Editions, 2011).
35. Having a specific identity – knowing who you are – shouldn't be confused with simply being self-conscious, namely, conscious that you are conscious (metarepresentation: literally, a 'higher order representation'). By the same token, an absence of metarepresentation is sometimes confused with autism. However, just because an autistic child cannot understand that someone might have different beliefs from them doesn't mean to say that they are not aware that they are aware, namely, that they are self-conscious – have metarepresentation. Von Eckardt, B. in *MIT Encyclopedia of Cognitive Science* (eds. Wilson, R. & Keil, F.) (MIT Press, 1999). Leslie, A. M. 'The theory of mind impairment in autism. Evidence for a modular mechanism of development?' in *Natural Theories of Mind: Evolution, Development and Simulation of Everyday Mindreading* (ed. Whiten, A.) (Blackwell, 1991).
36. Cleeremans, A. 'Consciousness: the radical plasticity thesis'. *Progress in Brain Research*, 168, 19–33 (2008).
37. Persaud, N. & McLeod, P. 'Wagering demonstrates subconscious processing in a binary exclusion task'. *Consciousness and Cognition*, 17, 565–75 (2008).

38. Greenfield, S. A. *The Private Life of the Brain: Emotions, Consciousness and the Secret of the Self.* (Wiley, 2000).
39. Dennett, D. C., 'Are we explaining consciousness yet?' *Cognition*, 79, 221–37 (2001).
40. Rosenthal, D. 'A theory of consciousness' in *The Nature of Consciousness: Philosophical Debates* (eds. Block, N., Flanagan, O. & Guzeldere, G.) (MIT Press, 1997).
41. Suh, E. M. 'Culture, identity consistency and subjective well-being'. *Journal of Personality and Social Psychology*, 83, 1378–91 (2002).
42. Martinsen, Ø. L. 'The creative personality: a synthesis and development of the creative person profile'. *Creativity Research Journal*, 23, 185–202 (2011).
43. Maddux, W. W., Adam, H. & Galinsky, A. D. 'When in Rome . . . Learn why the Romans do what they do: how multicultural learning experiences facilitate creativity'. *Personality and Social Psychology Bulletin*, 36, 731–41 (2010).
44. The Maddox et al. (2010) study does not control for various confounding factors. For example, an existing passion for travel could highlight innate curiosity and a particular intelligence that might also predict performance in the creativity task. The focus of the study seems to be learning in a foreign country, which the researchers contrast with learning a sport, to show that learning in a foreign country (and recalling the experience) improves creativity.
45. Thanem, Varlander & Cummings, 2011.
46. Brown, S. *Play: How It Shapes the Brain, Opens the Imagination and Invigorates the Soul.* (Avery, 2009).
47. Fleming, P. 'Workers' playtime? Boundaries and cynicism in a "culture of fun" program'. *Journal of Applied Behavioral Science*, 41, 285–303 (2005).
48. Leung, A. K. et al. 'Embodied metaphors and creative "acts"'. *Psychological Science*, 23, 502–9 (2012).
49. Mann, S. & Cadman, R. 'Does being bored make us more creative?' *Creativity Research Journal*, 26, 165–73 (2014).
50. Dijksterhuis, A. & Meurs, T. 'Where creativity resides: the generative power of unconscious thought'. *Consciousness and Cognition*, 15, 135–46 (2006).
51. Marshall, B. & Azad, M. 'Q&A: Barry Marshall. A bold experiment'. *Nature*, 514, S6–7 (2014).

52. Sass, L. A. 'Schizophrenia, modernism and the "creative imagination": on creativity and psychopathology'. *Creativity Research Journal*, 13, 55–74 (2001).

6. PROBLEMS AT HOME

1. Sturman, D. A. & Moghaddam, B. 'The neurobiology of adolescence: changes in brain architecture, functional dynamics and behavioral tendencies'. *Neuroscience and Biobehavioral Reviews*, 35, 1704–12 (2011).
2. Rivers, S. E., Reyna, V. F. & Mills, B. 'Risk-taking under the influence: a fuzzy-trace theory of emotion in adolescence'. *Developmental Review*, 28, 107–44 (2008).
3. Sturman & Moghaddam, 2011.
4. Fuster, J. *The Prefrontal Cortex*. (Academic Press, 2008).
5. Established knowledge indeed has it that the prefrontal cortex constituted some 30 per cent of the human brain, compared to only 17 per cent in our nearest relatives, the chimpanzees. However, recent work suggests a greater number of distributed networks are more vital to the unique abilities of humans and backs up an earlier study which found that white matter in the prefrontal cortex was greater in humans: Barton, R. A. & Venditti, C. 'Human frontal lobes are not relatively large'. *Proceedings of the National Academy of Sciences of the United States of America*, 110(22), 9001–6 (2013). Schoenemann, P. T., Sheehan, M. J. & Glotzer, L. D. 'Prefrontal white matter volume is disproportionately larger in humans than in other primates'. *Nature Neuroscience*, 8(2), 242–52 (2005). McBride, T., Arnold, S. E. & Gur, R. C. 'A comparative volumetric analysis of the prefrontal cortex in human and baboon MRI'. *Brain, Behavior and Evolution*, 54(3), 159–66 (1999).
6. Tsujimoto, S. 'The prefrontal cortex: functional neural development during early childhood'. *Neuroscientist*, 14, 345–58 (2008).
7. Alvarez, J. A. & Emory, E. 'Executive function and the frontal lobes: a meta-analytic review'. *Neuropsychological Review*, 16, 17–42 (2006).
8. Sturman & Moghaddam, 2011.
9. Casey, B. J., Getz, S. & Galvan, A. 'The adolescent brain'. *Developmental Review*, 28, 62–77 (2008).
10. Chambers, R. A., Taylor, J. R. & Potenza, M. N. 'Developmental neurocircuitry of motivation in adolescence: a critical period of addiction vulnerability'. *American Journal of Psychiatry*, 160, 1041–52 (2003).

11. Ferron, A. et al. 'Inhibitory influence of the mesocortical dopaminergic system on spontaneous activity or excitatory response induced from the thalamic mediodorsal nucleus in the rat medial prefrontal cortex'. *Brain Research*, 302, 257–65 (1984); see also Gao, W.-J., Wang, Y. & Goldman-Rakic, P. S. 'Dopamine modulation of perisomatic and peri-dendritic inhibition in prefrontal cortex'. *Journal of Neuroscience*, 23, 1622–30 (2003).
12. Knobloch, H. S. & Grinevich, V. 'Evolution of oxytocin pathways in the brain of vertebrates'. *Frontiers in Behavioral Neuroscience*, 8, 31 (2014).
13. Steinberg, L. 'A social neuroscience perspective on adolescent risk-taking'. *Developmental Review*, 28, 78–106 (2008).
14. Barch, D. M. 'The cognitive neuroscience of schizophrenia'. *Annual Review of Clinical Psychology*, 1, 321–53 (2005); see also Thoma, P. et al. 'Proverb comprehension impairments in schizophrenia are related to executive dysfunction'. *Psychiatry Research*, 170, 132–9 (2009).
15. Cortiñas, M. et al. 'Reduced novelty-P3 associated with increased behavioral distractibility in schizophrenia'. *Biological Psychology*, 78, 253–60 (2008).
16. Oltmanns, T. F. 'Selective attention in schizophrenic and manic psychoses: the effect of distraction on information processing'. *Journal of Abnormal Psychology*, 87, 212–25 (1978); see also Parsons, B. D. et al. 'Lengthened temporal integration in schizophrenia'. *Neuropsychologia*, 51, 372–6 (2013).
17. Strange, P. G. *Brain Biochemistry and Brain Disorders*. (Oxford University Press, 1992); Sturman & Moghaddam, 2011.
18. Evidence for this long-established idea is straightforward enough: a stimulant drug such as amphetamine, which enhances the release of dopamine in the brain, mimics much of the psychotic syndrome of schizophrenia. Conversely, antipsychotic drugs, which block the action of dopamine, have a major tranquillizing action and have long been used as the therapy of choice in schizophrenia. However, a word of caution: in schizophrenia itself, the excessive influence of dopamine may not be so much due to a literal abundance of the number of physical transmitter molecules, but to an amplification of the effects of an otherwise normal amount of dopamine via an abnormality in their target receptors, or some other aberrant brain process. In any event, it is extremely unlikely that such a range of complex sensory and cognitive impairments such as those which characterize schizophrenia could be attributable simply to one transmitter system. The key point here, nonetheless, is that dopamine plays an important part – albeit not an

exclusive one – in the delicate interplay of brain chemicals, their targets and the consequent reconfiguring of neuronal circuits that eventually makes up the individual 'mind'. If so, and if, as suggested earlier, 'understanding' is seeing one thing in terms of another, then a poverty of connections will lead to a lack of understanding, especially when it comes to such abstracted concepts as those embodied in proverbs. Moreover, and more generally, both children and schizophrenics seem to lack logic and their thought processes are often characterized by a fragile, idiosyncratic rationale. These seemingly irrational and surprising associations would be apparent in any of the wide range of vehicles available for self-expression, be it the odd, unrealistic combinations of colours in a child's painting, such as a sheep coloured purple, or the abstracted visual patterns of jumbled 'word salads', an unintelligible mixture of words or phrases, so common in a child's art or a schizophrenic's poetry. Brisch, R. et al. 'The role of dopamine in schizophrenia from a neurobiological and evolutionary perspective: old-fashioned, but still in vogue'. *Frontiers in Psychiatry*, 5, 47 (2014). Kasanin, J. S. *Language and Thought in Schizophrenia*. (University of California Press, 1944). Mujica-Parodi, L. R., Malaspina, D. & Sackeim, H. A. 'Logical processing, affect and delusional thought in schizophrenia'. *Harvard Review of Psychiatry*, 8, 73–83. Caplan, R. et al. 'Formal thought disorder in childhood onset schizophrenia and schizotypal personality disorder'. *Journal of Child Psychology and Psychiatry*, 31, 1103–14 (1990).

19. Gao, Wang & Goldman-Rakic, 2003.
20. Tsujimoto, 2008; see also Welsh, M. C. & Pennington, B. F. 'Assessing frontal lobe functioning in children: views from developmental psychology. *Developmental Neuropsychology*, 4, 199–230 (1988).
21. Parsons et al., 2013.
22. Callicott, J. H. et al. 'Physiological dysfunction of the dorsolateral prefrontal cortex in schizophrenia revisited'. *Cerebral Cortex*, 10, 1078–92 (2000).
23. Ferron et al., 1984.
24. Davis, C. et al. 'Decision-making deficits and overeating: a risk model for obesity'. *Obesity Research & Clinical Practice*, 12, 929–35 (2004); see also Pignatti, R. et al. 'Decision-making in obesity: a study using the Gambling Task'. *Eating and Weight Disorders*, 11, 126–32 (2006).
25. Tataranni, P. A. & DelParigi, A. 'Functional neuroimaging: a new generation of human brain studies in obesity research'. *Obesity Reviews*, 4, 229–38 (2003).

26. Tanabe, J. et al. 'Prefrontal cortex activity is reduced in gambling and non-gambling substance users during decision-making'. *Human Brain Mapping*, 28, 1276–86 (2007).
27. Shimamura, A. P. 'Memory and the prefrontal cortex'. *Annals of the New York Academy of Sciences*, 769, 151–9 (1995).
28. Cole, M. W. et al. 'Global connectivity of prefrontal cortex predicts cognitive control and intelligence'. *Journal of Neuroscience*, 32, 8988–99 (2012).
29. The table is adapted from one in Greenfield, S. A. *Mind Change: How Digital Technologies are Leaving Their Mark on Our Brains.* (Random House, 2014).
30. Rosen, L. D. et al. 'Media and technology use predicts ill-being among children, preteens and teenagers independent of the negative health impacts of exercise and eating habits'. *Computers in Human Behavior*, 35, 364–75 (2014).
31. Professor Michael Merzenich, from the University of California, San Francisco, and an expert in the experience-induced plasticity of the brain, gives the corresponding typical neuroscientific perspective. He cautions, 'There is a massive and unprecedented difference in how their [the digital natives'] brains are plastically engaged in life, compared with those of average individuals from earlier generations, and there is little question that the operational characteristics of the average modern brain substantially differ.' The evidence for such changes and the effects on the way the next generation may think and feel, along with the evidence already accumulating of these effects, have now already been documented: suffice it to say that this radical shift into a cyber-environment might lead in general to a profile of more agile mental processing and yet an inappropriate degree of recklessness, poor empathy and interpersonal skills, a narcissistic predisposition, low-grade aggression and a weak sense of identity, which can be set next to the benefits of improved sensory motor coordination, possible better performance on IQ tests, fast reaction times and improved working memory. Bavelier, D. et al. 'Brains on video games'. *National Review of Neuroscience*, 12, 763–8 (2011). Greenfield, S. A., 2014.
32. Koepp, M. J. et al. 'Evidence for striatal dopamine release during a video game'. *Nature*, 393, 266–8 (1998); see also Weinstein, A. M. 'Computer and video game addiction – a comparison between game users and non-game users'. *American Journal of Drug and Alcohol Abuse*, 36, 268–76 (2010).
33. Greenfield, S. A., 2014.

34. Yuan, K. et al. 'Microstructure abnormalities in adolescents with internet addiction disorder'. *PLoS One*, 6, e20708 (2011).
35. This is a controversial and complex topic that is fully discussed in Greenfield, S. A., 2014.
36. Freis, E. D. & Ari, R. 'Clinical and experimental effects of reserpine in patients with essential hypertension'. *Annals of the New York Academy of Sciences*, 59, 45–53 (1954).
37. Nutt, D. J. 'The role of dopamine and norepinephrine in depression and antidepressant treatment'. *Journal of Clinical Psychiatry*, 67 (suppl. 6), 3–8 (2006).
38. Pletscher, A. 'The discovery of antidepressants: a winding path'. *Experientia*, 47, 4–8 (1991).
39. Healy, D. *The Antidepressant Era*. (Harvard University Press, 1997).
40. Since that time, needless to say, more pharmaceutically sophisticated and selective drugs have been developed. For example, the prototype MAOIs proved to have undesirable side effects, such as the 'cheese reaction', whereby anyone taking them would have to avoid foods rich in the chemical tyramine, such as cheese, Marmite, chocolate or red wine: the now drug-protected, non-broken-down tyramine (another cousin of noradrenaline) would thus then accumulate, reaching levels that consequently raised blood pressure and heart rate and increasing the risk of hypertensive crisis. Accordingly, a new class of tricyclic drugs, so called because of their three-ringed chemical structure, was developed which avoided this problem by simply targeting the amine transmitters more selectively. These newer drugs worked by preventing the re-uptake of noradrenaline and dopamine into neurons, leaving them available therefore to have a longer action on their various targets: the tyramine in foodstuffs was now unaffected and patients on tricyclics could once again enjoy cheese and wine. More recently still, antidepressant medication has become still further refined, with the third generation of drugs such as Prozac, which are more selective still, this time primarily for the amine serotonin.
41. Fitzgerald, P. J. 'Forbearance for fluoxetine: do monoaminergic antidepressants require a number of years to reach maximum therapeutic effect in humans?' *International Journal of Neuroscience*, 124, 467–73 (2014).
42. Scott, J. 'Cognitive therapy'. *British Journal of Psychiatry*, 165, 126–30 (1994); see also Cuijpers, P. et al. 'A meta-analysis of cognitive-behavioural therapy for adult depression, alone and in comparison with other treatments'. *Canadian Journal of Psychiatry*, 58, 376–85 (2013).

43. Anacker, C. 'Adult hippocampal neurogenesis in depression: behavioral implications and regulation by the stress system'. *Current Topics in Behavioral Neurosciences* (2014).
44. Sheline, Y. I. et al. 'Resting-state functional MRI in depression unmasks increased connectivity between networks via the dorsal nexus'. *Proceedings of the National Academy of Sciences of the United States of America*, 107, 11020–5 (2010).
45. Ren, J. et al. 'Repetitive transcranial magnetic stimulation versus electroconvulsive therapy for major depression: a systematic review and meta-analysis'. *Progress in Neuropsychopharmacology & Biological Psychiatry*, 51, 181–9 (2014).
46. The efficacy of lithium was first reported in 1948 by the Australian physician John Cade. Once again, serendipity played a key role: Cade was exploring the possible biochemical bases of mental disorders by injecting urine from mentally ill patients into the abdomen of guinea pigs. The animals died faster than when urine samples from healthy people were used: hence Cade concluded that more uric acid might be present in the samples provided by his mentally ill patients. It was then that fate played a hand: as Cade tried to increase the water solubility of uric acid by adding lithium urate to the solution, he found that in the guinea pigs injected with the lithium urate solution, toxicity was indeed greatly reduced. However, the more fascinating result was that the lithium had by itself a calming effect on the animals. Why? The powerful therapeutic effects of lithium, specifically in manic depression, as opposed to steady-state unipolar depression, have proved a riddle. Needless to say, there are various theories, including an action on the amino acid glutamate, itself an 'excitatory' transmitter and/or a modulatory enhancement of serotonin. Another target might be the smallest transmitter of all, the gas nitric oxide, which is implicated in plasticity. Yet other ideas are that lithium could somehow reset the body clock regulating metabolism, temperature and sleep (disrupted in bipolar disorder), or inhibit an enzyme (inositol monophosphotase) which would otherwise impair a chemical (inositol), aberrations in which lead to memory problems – and depression. However, the problem with all these sophisticated possible mechanisms is that, in normal, healthy people, lithium has no psychoactive effect at all and, moreover, we're talking here about nothing more than a simple salt. How might something so basic have any kind of selective, sophisticated effect on such a complex mental disorder? Yoshimura, R. et al. 'Comparison of lithium, aripiprazole and olanzapine as augmentation to paroxetine

for inpatients with major depressive disorder'. *Therapeutic Advances in Psychopharmacology*, 4, 123–9 (2014). De Sousa, R. T. et al. 'Lithium increases nitric oxide levels in subjects with bipolar disorder during depressive episodes'. *Journal of Psychiatric Research*, 55, 96–100 (2014). Welsh, D. K. & Moore-Ede, M. C. 'Lithium lengthens circadian period in a diurnal primate, Saimiri sciureus'. *Biological Psychiatry*, 28, 117–26 (1990). Brown, K. M. & Tracy, D. K. 'Lithium: the pharmacodynamic actions of the amazing ion'. *Therapeutic Advances in Psychopharmacology*, 3, 163–76 (2013). Buigues, C. et al. 'The relationship between depression and frailty syndrome: a systematic review'. *Aging & Mental Health*, 1–11 (16 Oct. 2014).

47. Yanagita, T. et al. 'Lithium inhibits function of voltage-dependent sodium channels and catecholamine secretion independent of glycogen synthase kinase-3 in adrenal chromaffin cells'. *Neuropharmacology*, 53, 881–9 (2007).
48. Mason, L. et al. 'Decision-making and trait impulsivity in bipolar disorder are associated with reduced prefrontal regulation of striatal reward valuation'. *Brain*, 137, 2346–55 (2014).
49. Hercher, C., Chopra, V. & Beasley, C. L. 'Evidence for morphological alterations in prefrontal white matter glia in schizophrenia and bipolar disorder'. *Journal of Psychiatry and Neuroscience*, 39, 130277 (2014).
50. Muzina, D. J. & Calabrese, J. R. 'Maintenance therapies in bipolar disorder: focus on randomized controlled trials'. *Australian and New Zealand Journal of Psychiatry*, 39, 652–61 (2005).
51. Lautenbacher, S. & Krieg, J. C. 'Pain perception in psychiatric disorders: a review of the literature'. *Journal of Psychiatric Research*, 28, 109–22 (1994); see also Guieu, R., Samuélian, J. C. & Coulouvrat, H. 'Objective evaluation of pain perception in patients with schizophrenia'. *British Journal of Psychiatry*, 164, 253–5 (1994); and Dworkin, R. H. 'Pain insensitivity in schizophrenia: a neglected phenomenon and some implications'. *Schizophrenia Bulletin*, 20, 235–48 (1994).
52. Giles, L. L., Singh, M. K. & Nasrallah, H. A. 'Too much or too little pain: the dichotomy of pain sensitivity in psychotic versus other psychiatric disorders'. *Current Psychosis and Therapeutics Reports*, 4, 134–8 (2006).
53. Adler, G. & Gattaz, W. F. 'Pain perception threshold in major depression'. *Biological Psychiatry*, 34, 687–9 (1993); see also Diener, H. C., van Schayck, R. & Kastrup, O. 'Pain and depression' in *Pain and the Brain from Nociception to Cognition* (eds. Bromm, B. & Desmedt, J. E.) 345–55 (Raven Press, 1995).

54. Then again, Jasper and Penfield found that removing quite large areas of the cortex (and different areas) in various epileptic patients didn't stop people experiencing pain. Penfield, W. & Jasper, H. 'Epilepsy and the functional anatomy of the human brain'. (Little, Brown & Co., 1954).
55. Hall, K. R. & Stride, E. 'The varying response to pain in psychiatric disorders: a study in abnormal psychology'. *British Journal of Medical Psychology*, 27, 48–60 (1954); see also Pöllmann, L. & Harris, P. H. 'Rhythmic changes in pain sensitivity in teeth'. *International Journal of Chronobiology*, 5, 459–64 (1978).
56. Koyama, T. et al. 'The subjective experience of pain: where expectations become reality. *Proceedings of the National Academy of Sciences of the United States of America*, 102, 12950–5 (2005).
57. Ramachandran, V. S. & Blakeslee, S. *Phantoms in the Brain: Probing the Mysteries of the Human Mind.* (Harper Perennial, 1998).
58. Melzack, R. & Wall, P. D. *The Challenge of Pain.* (Penguin, 1996).
59. Kupers, R. C., Konings, H., Adriaensen, H. & Gybels, J. M. 'Morphine differentially affects the sensory and affective pain ratings in neurogenic and idiopathic forms of pain'. *Pain*, 47, 5–12 (1991).
60. Ponterio, G. et al. 'Powerful inhibitory action of mu opioid receptors (MOR) on cholinergic interneuron excitability in the dorsal striatum'. *Neuropharmacology*, 75, 78–85 (2013).
61. Lautenbacher & Krieg, 1994; see also Guieu, Samuélian & Coulouvrat, 1994; and Dworkin, 1994.
62. Rang, H. P. et al. *Rang and Dale's Pharmacology.* (Churchill Livingstone, 2012).
63. Bergman, N. A. 'Michael Faraday and his contribution to anesthesia'. *Anesthesiology*, 77, 812–16 (1992).
64. http://www.dailymail.co.uk/news/article-2377857/Nitrous-oxide-Laughing-gas-known-hippy-crack-2nd-popular-legal-high-drug-young-people.html
65. Featured in the news recently: http://www.theguardian.com/world/2015/feb/27/raver-drug-ketamine-control-plan-at-un-condemned-as-potential-disaster
66. Buigues, C. et al. 'The relationship between depression and frailty syndrome: a systematic review'. *Aging & Mental Health*, 1–11 (2014).
67. Winocur, G. & Moscovitch, M. 'A comparison of cognitive function in community-dwelling and institutionalized old people of normal intelligence'. *Canadian Journal of Psychology*, 44, 435–44 (1990).
68. Scarmeas, N. & Stern, Y. 'Cognitive reserve and lifestyle'. *Journal of Clinical and Experimental Neuropsychology*, 25, 625–33 (2003).

69. Engvig, A. et al. 'Effects of memory training on cortical thickness in the elderly'. *NeuroImage*, 52, 1667–76 (2010).
70. Small, G. W. & Greenfield, S. 'Current and future treatments for Alzheimer disease'. *American Journal of Geriatric Psychiatry*, 23, 1101–5 (2015).
71. One popular idea, currently at the heart of much research into new, more effective medication, is that the chemical culprit really responsible for the neuronal devastation in Alzheimer's is an abnormally formed substance named after the Greek for 'starch': amyloid. However, as yet, drugs combating the presence of amyloid have been disappointing in clinical trials held for over the course of a decade. It seems that these abnormal deposits in the brain may be an adjunct factor but yet, once again, not the central problem: amyloid can be present in erstwhile healthy brains post mortem, so, at most, its presence will be a necessary, not a sufficient factor. An alternative target currently being investigated is another histological abnormality found in Alzheimer brains: 'tangles', a misarrangement of the microtubules we first encountered back in Chapter 1. There is no reason why amyloid 'plaques' or tau 'tangles' should be the initiating cause of the cell death, and hence the diminution of the assembly, in Alzheimer's disease. Moreover, plaques and tangles would be non-specific killers of all cells: see Small & Greenfield, ibid. in contrast, the primary culprit for Alzheimer's must be different and much more selective, as only certain neuron groups in the brain are vulnerable to degeneration. Greenfield, S. A. & Vaux, D. J. 'Parkinson's disease, Alzheimer's disease and motor neurone disease. Identifying a common mechanism'. *Neuroscience*, 113, 485–92 (2002).
72. Aarsland, D. et al. 'Frequency of dementia in Parkinson disease'. *Archives of Neurology*, 53, 538–42 (1996); see also Calne, D. B. et al. 'Alzheimer's disease, Parkinson's disease and motorneurone disease: abiotrophic interaction between ageing and environment?' *Lancet*, 2, 1067–70 (1986); and Horvath, J. et al. 'Neuropathology of Parkinsonism in patients with pure Alzheimer's disease'. *Journal of Alzheimer's Disease*, 39, 115–20 (2014).
73. Whenever most brain cells are damaged – say, by a stroke or blow to the head – recovery of function will normally occur to some extent. However, if the damage happens to be to the 'hub' cells, this triggers the still-present developmental mechanism: the release of a particular chemical that is only usually operative in the developing brain. Things are different, however, in the mature brain: such developmental agents turn

out now to be toxic. So the chemical that was released as compensation for the initially damaged cells will now generate still more damage, and therefore cause still more release of the now-toxic chemical. This is the cycle of cell death we see as neurodegeneration. Greenfield, S. 'Discovering and targeting the basic mechanism of neurodegeneration: the role of peptides from the C-terminus of acetylcholinesterase: non-hydrolytic effects of ache: the actions of peptides derived from the C-terminal and their relevance to neurodegeneration'. *Chemico-biological Interactions*, 203, 543–6 (2013).

74. Woods, B. et al. 'Reminiscence therapy for dementia'. *Cochrane Database of Systematic Reviews* CD001120 (2005).
75. Atkins, S. *First Steps: Living with Dementia.* (Lion, 2013).
76. Sacks, O. *Musicophilia: Tales of Music and the Brain.* (Random House, 2007).
77. Yet another approach for offsetting the decline in checks and balances and hence the immersion in 'booming, buzzing confusion' is Reality Orientation. This type of therapy aims to reduce confusion in people with dementia. The strategy is to orientate the person to the place and time they are in currently and to the names, identities and caring roles of the people around them. In short, this approach tries to allow people to know where and who they and those around them are, so as to reduce uncertainty and anxiety. Examples of this approach would be having a board on display with the day, date, the next meal and the weather written on it, mounting a large calendar clock on the wall and displaying the names of rooms on doors with the person's own name as the label to their bedroom. In a sense, the normal context of time and space in daily life which would normally be provided by internal neuronal connectivity is now supplied externally in the outside environment. One problem with this approach could be that being corrected about situations can actually make Alzheimer patients feel worse. For example, telling someone they can't go home to their husband because he died ten years ago can be very distressing for someone who has forgotten about the death and may therefore react to the news as they would the first time they heard it. However, the crucial issue is to let the person guide the interpersonal interaction, in a sense, to determine the 'reality'. No halfway right-minded adult would tell a small child that Father Christmas did not exist; rather, they would enter into the childhood narrative, paying deference to the child's limited grasp of reality, and indeed adopt different perspectives to that of a grown-up. This was the approach described by psychologist Oliver James, writing in the

Guardian in 2008 about the way in which a daughter, Penny Garner, developed a means for coping with her mother's dementia: 'Garner's ideas evolved as a result of caring for her mother, Dorothy Johnson, when she developed Alzheimer's. One day, they were sitting together in a doctor's waiting room when out of the blue Dorothy said, "Has our flight been called yet?" Garner was mystified and played for time. Her mother anxiously looked around and said, "We don't want to miss it, where's our hand luggage?" Suddenly, Garner realized what was happening. Her mother had always loved air travel and Dorothy was making sense of this crowded waiting situation by assuming they were in a departure lounge. When Garner responded with "All our luggage has been checked in, we've just got our handbags", her mother visibly relaxed.' Here, it's not so much trying to increase the stone with supernumerary connections, nor give chemicals that will help improve the viscosity of the puddle, but rather manipulating the outside environment, conversation and interaction, so that the existing internal neuronal connectivity is appropriate. Atkins, 2013. Spector, A. et al. 'Reality orientation for dementia'. *Cochrane Database of Systematic Reviews* CD001119 (2000).

7. DREAMING

1. Capellini, I. et al. 'Phylogenetic analysis of the ecology and evolution of mammalian sleep'. *Evolution*, 62, 1764–6 (2008).
2. Foulkes, D. 'Dreaming and REM sleep'. *Journal of Sleep Research*, 2, 199–202 (1993).
3. Nir, Y. & Tononi, G. 'Dreaming and the brain: from phenomenology to neurophysiology'. *Trends in Cognitive Science*, 14, 88–100 (2010).
4. Morris, G. O., Williams, H. L. & Lubin, A. 'Misperception and disorientation during sleep deprivation'. *Archives of General Psychiatry*, 2, 247–54 (1960).
5. Harrison, Y. & Horne, J. A. 'Sleep loss and temporal memory'. *Quarterly Journal of Experimental Psychology. A.*, 53, 271–9 (2000).
6. Zohar, D. et al. 'The effects of sleep loss on medical residents' emotional reactions to work events: a cognitive-energy model'. *Sleep*, 28, 47–54 (2005).
7. Yoo, S.-S. et al. 'The human emotional brain without sleep – a prefrontal amygdala disconnect'. *Current Biology*, 17, R877–8 (2007).
8. van der Helm, E. et al. 'REM sleep depotentiates amygdala activity to previous emotional experiences'. *Current Biology*, 21, 2029–32 (2011).

9. Walker, M. P. 'Why we sleep?' in Brain.org Gulbenkian Health Forum (2012).
10. Hobson, J. A. & Friston, K. J. 'Waking and dreaming consciousness: neurobiological and functional considerations'. *Progress in Neurobiology*, 98, 82–98 (2012).
11. It is unclear whether the idea of the input from the senses would include taste and smell, which seem to feature far less in dreams.
12. Hobson & Friston, 2012.
13. Mignot, E. 'Why we sleep: the temporal organization of recovery'. *PLoS Biology*, 6, e106 (2008).
14. Hobson & Friston, 2012.
15. Marchant, J. 'Why brainy animals need more REM sleep after all'. *New Scientist* (19 June 2008).
16. Siegel, J. M. 'The evolution of REM sleep' in *Handbook of Behavioral State Control* (eds. Lydic, R. & Baghdoyan, H. A.) (CRC Press, 1999).
17. Hobson, J. A. 'REM sleep and dreaming: towards a theory of protoconsciousness'. *National Review of Neuroscience*, 10, 803–13 (2009).
18. Davis, K. F., Parker, K. P. & Montgomery, G. L. 'Sleep in infants and young children: part one: normal sleep'. *Journal of Pediatric Health Care*, 18, 65–71 (2004).
19. http://www.babble.com/baby/baby-sleep-tips/baby-sleep-tips-1/www.babble.com/baby/baby-sleep/
20. Hobson & Friston, 2012.
21. Though NREM sleep could well play a significant role in memory consolidation: O'Neill J. et al. 'Play it again: reactivation of waking experience and memory'. *Trends in Neuroscience*, 33(5), 220–9 (2010).
22. Kumar, S. & Sagili, H. 'Etiopathogenesis and neurobiology of narcolepsy: a review'. *Journal of Clinical and Diagnostic Research*, 8, 190–5 (2014).
23. Morrison, A. R. 'A window on the sleeping brain' in *The Workings of the Brain: Development, Memory and Perception* (ed. Llinas, R. R.), 133–148 (W. H. Freeman, 1990).
24. Nir & Tononi, 2010.
25. Llinás, R. R. & Paré, D. 'Of dreaming and wakefulness'. *Neuroscience*, 44, 521–35 (1991).
26. Many riddles would remain in the Llinás and Paré scenario, such as how to account for the enormous variations in levels of consciousness across species and in the foetus: yet these wide differentials would not be apparent from the simple thalamocortical dynamics common across species, not to mention the consciousness-modifying properties of

various psychoactive drugs on within-individual fluctuations in levels of consciousness. Incidentally, this is a flaw also in more recent enthusiasm for the thalamocortical loop by Tononi and Koch, who speak of a 'bistability' in thalamocortical networks, implying that consciousness is all or nothing, and therefore operated by some kind of neuronal on–off switch. Tononi, G. & Koch, C. 'The neural correlates of consciousness: an update'. *Annals of the New York Academy of Sciences*, 1124, 239–61 (2008).

27. Plum, F. in 'Coma and related disturbances of the human conscious state'. *Cerebral Cortex Vol. 9: Normal and Altered States and Function* (eds. Peters, A. & Jones, E. G.) 359–426 (Plenum Press, 1991); see also Young, G. B., Ropper, A. H. & Bolton, C. E. *Coma and Impaired Consciousness: A Clinical Perspective*. (McGraw Hill, 1998).
28. The area of the parieto-occipital-temporal (POT) junction includes portions of the parietal, temporal and occipital lobes and serves as a location for integration of auditory, visual and somatosensory inputs, while sending outputs to a wide range of other areas, including the limbic areas and the prefrontal cortex. See also: Maquet, P. et al. 'Functional neuroanatomy of human rapid-eye-movement sleep and dreaming'. *Nature*, 383, 163–6 (1996).
29. Epstein, A. W. 'Effect of certain cerebral hemispheric diseases on dreaming'. *Biological Psychiatry*, 14, 77–93 (1979); see also Maquet, P. 'Functional neuroimaging of normal human sleep by positron emission tomography'. *Journal of Sleep Research*, 9, 207–31 (2000); and Jakobson, A. J., Fitzgerald, P. B. & Conduit, R. 'Induction of visual dream reports after transcranial direct current stimulation (tDCs) during Stage 2 sleep'. *Journal of Sleep Research*, 21, 369–79 (2012).
30. Solms, M. 'Dreaming and REM sleep are controlled by different brain mechanisms'. *Behavioral and Brain Sciences*, 23, 843–50; discussion 904–1121 (2000).
31. Büchel, C. et al. 'Different activation patterns in the visual cortex of late and congenitally blind subjects'. *Brain*, 121, Pt 3, 409–19 (1998).
32. Ibid.
33. Horikawa, T. et al. 'Neural decoding of visual imagery during sleep'. *Science*, 340, 639–42 (2013).
34. Nir & Tononi, 2010.
35. Ibid.
36. Penfield, W. & Perot, P. 'The brain's record of auditory and visual experience. A final summary and discussion'. *Brain*, 86, 595–696 (1963).

37. Hobson, J. A. *The Chemistry of Conscious States: How the Brain Changes Its Mind.* (Little, Brown & Co., 1994); see also Scarone, S. et al. 'The dream as a model for psychosis: an experimental approach using bizarreness as a cognitive marker'. *Schizophrenia Bulletin*, 34, 515–22 (2008); and Limosani, I. et al. 'The dreaming brain/mind, consciousness and psychosis'. *Consciousness and Cognition*, 20, 987–92 (2011).
38. Brisch, R. et al. 'The role of dopamine in schizophrenia from a neurobiological and evolutionary perspective: old-fashioned, but still in vogue'. *Frontiers in Psychiatry*, 5, 47 (2014).
39. Solms, M. & Turnbull, O. *The Brain and the Inner World.* (Other Press, 2002).
40. Monti, J. M. & Monti, D. 'The involvement of dopamine in the modulation of sleep and waking'. *Sleep Medicine Reviews*, 11, 113–33 (2007).
41. Normally, the dopamine fountain can be regulated by a transmitter closely linked to REM sleep, namely, by acetylcholine (ACh) operating deep in the brain, in the brain stem: however, dopamine can also go to work on the cortex independently of these ACh systems when they are not operational, as happens in dreaming when it occurs independently of REM. This time, the dopamine is working further up the chain of command, beyond the REM-generating brainstem mechanisms and more directly on the final processes in the prefrontal cortex which correlate with actual dreaming itself. So, then, what would cause the dopamine to be released in this case? Nothing. It turns out that nothing further at all would be required, not even sensory inputs. In the absence of inputs from eyes, ears, and so on, and when acetylcholine is not acting as a trigger, as is the case with dreaming in non-REM, the dopamine cells in the brain stem display an 'autorhythmicity: a continuing cycle of varying excitability leading to periodic bursts of action potentials that in turn would lead to the release of dopamine in the 'higher' target brain regions. Grace, A. A. & Bunney, B. S. 'The control of firing pattern in nigral dopamine neurons: burst firing'. *Journal of Neuroscience*, 4, 2877–90 (1984). Grace, A. A. & Bunney, B. S. 'The control of firing pattern in nigral dopamine neurons: single spike firing'. *Journal of Neuroscience*, 4, 2866–76 (1984).
42. Ferron, A. et al. 'Inhibitory influence of the mesocortical dopaminergic system on spontaneous activity or excitatory response induced from the thalamic mediodorsal nucleus in the rat medial prefrontal cortex'. *Brain Research*, 302, 257–65 (1984); see also Gao, W.-J., Wang, Y. & Goldman-Rakic, P. S. 'Dopamine modulation of perisomatic and

peridendritic inhibition in prefrontal cortex'. *Journal of Neuroscience*, 23, 1622–30 (2003).

43. Nir & Tononi, 2010; see also Dang-Vu, T. T. et al. 'Functional neuroimaging insights into the physiology of human sleep'. *Sleep*, 33, 1589–603 (2010).
44. Solms, 2000.
45. Firstly, within different parts of the neuron, dopamine might be having other effects on more subtle local currents and hence activities that don't necessarily contribute at all to action potentials, and would therefore not register with the recording methods used here. Secondly, within the prefrontal cortex the inhibition that would result from the sporadic release of dopamine could have a net effect of disinhibition on connecting cells within local circuits: this action would then change the final output from the prefrontal cortex itself in a way that the complete and blanket absence of activity could not do. Thirdly, a large experimental lesion of the prefrontal cortex would result in all cells being destroyed indiscriminately, whereas the selective natural targeting of the dopamine in the normal, intact brain would inevitably not impact on all prefrontal cells equally. Lambert, R. C. et al. 'The many faces of T-type calcium channels'. *Pflügers Archiv. European Journal of Physiology*, 466, 415–23 (2014). Holthoff, K., Kovalchuk, Y. & Konnerth, A. 'Dendritic spikes and activity-dependent synaptic plasticity'. *Cell Tissue Research*, 326, 369–77 (2006).
46. Nielsen, T. A. et al. 'Pain in dreams'. *Sleep*, 16, 490–8 (1993).
47. Lautenbacher, S. & Krieg, J. C. 'Pain perception in psychiatric disorders: a review of the literature'. *Journal of Psychiatric Research*, 28, 109–22 (1994); see also Guieu, R., Samuélian, J. C. & Coulouvrat, H. 'Objective evaluation of pain perception in patients with schizophrenia'. *British Journal of Psychiatry*, 164, 253–5 (1994).
48. Casey, B. J., Getz, S. & Galvan, A. 'The adolescent brain'. *Developmental Review*, 28, 62–77 (2008); see also Sturman, D. A. & Moghaddam, B. 'The neurobiology of adolescence: changes in brain architecture, functional dynamics and behavioral tendencies'. *Neuroscience and Biobehavioral Reviews*, 35, 1704–12 (2011).
49. Toda, M. & Abi-Dargham, A. 'Dopamine hypothesis of schizophrenia: making sense of it all'. *Current Psychiatry Reports*, 9, 329–36 (2007).
50. Zadra, A. L., Nielsen, T. A. & Donderi, D. C. 'Prevalence of auditory, olfactory and gustatory experiences in home dreams'. *Perceptual and Motor Skills*, 87, 819–26 (1998).

51. Dopamine (or, rather, its metabolite, HVA) levels in cerebrospinal fluid (the fluid bathing the brain and spinal cord) increase gradually with age, while the 5HIAA metabolite of 5HT don't show any changes (comparing premature and term neonates). Pollin R. A. & Abman, S. H., *Fetal and Neonatal Physiology: Expert Consult*, vol. 2, 1777.
52. Klemm, W. R. 'Why does REM sleep occur? A wake-up hypothesis'. *Frontiers in Systems Neuroscience*, 5, 73 (2011).
53. Aserinsky, E. & Kleitman, N. 'Regularly occurring periods of eye motility, and concomitant phenomena, during sleep'. *Science*, 118, 273–4 (1953).
54. Mavromatis, A. *Hypnogogia: The Unique State of Consciousness between Wakefulness and Sleep*. (Routledge, Chapman and Hall, 1987).
55. LaBerge, S. & Levitan, L. 'Validity established of DreamLight cues for eliciting lucid dreaming'. *Dreaming*, 5, 159–168 (1995).

8. OVERNIGHT

1. Greenfield, S. A. *The Private Life of the Brain: Emotions, Consciousness and the Secret of the Self*. (Wiley, 2000).
2. Popper, K. *Conjectures and Refutations: The Growth of Scientific Knowledge*. (Routledge and Kegan Paul, 1963).
3. Bryan, A. et al. 'Functional Electrical Impedance Tomography by Evoked Response: a new device for the study of human brain function during anaesthesia'. *Proceedings of the Anaesthetic Research Society Meeting*, 428–9 (2010).
4. In fEITER, a small high-frequency electric current is passed through a series of electrodes fitted around the head and the resultant voltages are measured via an interspersed series of references. The ensuing resistance measurements are then used to construct a cross-sectional image of the electrical impedance within the brain, reflecting the electrical activity in different areas. The response of fEITER (less than a tenth of a millisecond) is so much faster than conventional brain imaging that evoked responses to external sound, light or touch can be captured in rapid succession as different areas of the brain respond, thus tracking how and where the brain is processing the incoming stimuli.
5. de Lange, C. '3D movie reveals how brain loses consciousness'. *New Scientist*, 14 June (2011).
6. Damasio, A. *Descartes' Error: Emotion, Reason and the Human Brain*. (Penguin, 2005).
7. Ibid.

8. Mayer, E. A. 'Gut feelings: the emerging biology of gut–brain communication'. *Nature Reviews Neuroscience*, 12, 453–66 (2011).
9. The insula, the anterior cingulate cortex, orbitofrontal cortex and amygdala, as well as deeper areas such as the hypothalamus and periaqueductal gray.
10. But how would the all-important bioactive peptides themselves, released outside the brain from these diverse parts of the body, be able to penetrate the inner cerebral sanctum to play their own part in the subsequent dynamics of assemblies? It turns out that circulating chemicals can impact directly on the brain by an interaction on special brain regions (such as the area postrema) located on the walls of the interconnecting, fluid-filled cavities within the brain (ventricles) which have been formed naturally during embryogenesis. Alternatively, many bioactive agents can invade the brain even more directly, by penetrating the blood–brain barrier: as its name suggests, this is a tightly knit wall of specialized cells which normally ensures that only small, or highly fat-soluble, molecules can pass readily from the blood into the brain. Another possibility is that compounds might access the brain from the periphery via the immune cells of the brain (microglia) as they sense any perturbations in the body environment. Critchley, H. D. & Harrison, N. A. 'Visceral influences on brain and behavior'. *Neuron*, 77, 624–38 (2013).
11. Ramirez, J. M. & Cabanac, M. 'Pleasure, the common currency of emotions'. *Annals of the New York Academy of Sciences*, 1000, 293–5 (2003).
12. Leonard, B. E. 'The immune system, depression and the action of antidepressants'. *Progress in Neuro-psychopharmacology and Biological Psychiatry*, 25, 767–80 (2001).
13. Martin, P. *The Sickening Mind: Brain, Behaviour, Immunity and Disease*. (HarperCollins, 1997).
14. Mykletun, A. et al. 'Levels of anxiety and depression as predictors of mortality: the HUNT study'. *British Journal of Psychiatry*, 195, 118–25 (2009).
15. Wulsin, L. R. & Singal, B. M. 'Do depressive symptoms increase the risk for the onset of coronary disease? A systematic quantitative review'. *Psychosomatic Medicine*, 65, 201–10 (2003).
16. Ader, R. & Cohen, N. 'Behaviorally conditioned immunosuppression and murine systemic lupus erythematosus'. *Science*, 215, 1534–6 (1982); see also Maier, S. F., Watkins, L. R. & Fleshner, M. 'Psychoneuroimmunology. The interface between behavior, brain and immunity'. *American Psychologist*, 49, 1004–17 (1994).

17. Hökfelt, T. et al. 'Coexistence of peptides and putative transmitters in neurons'. *Advances in Biochemical Psychopharmacology*, 22, 1–23 (1980).
18. Hökfelt, T. et al. Neuropeptides – an overview. *Neuropharmacology*, 39, 1337–56 (2000); see also Hökfelt, T. et al. 'Some aspects on the anatomy and function of central cholecystokinin systems'. *BMC Pharmacology and Toxicology*, 91, 382–6 (2002); and Hökfelt, T., Bartfai, T. & Bloom, F. 'Neuropeptides: opportunities for drug discovery'. *Lancet Neurology*, 2, 463–72 (2003); and Hökfelt, T., Pernow, B. & Wahren, J. 'Substance P: a pioneer amongst neuropeptides'. *Journal of Internal Medicine*, 249, 27–40 (2001).
19. Most articles are reluctant to give an actual number. Good recent reviews include Leng, G. & Ludwig, M. 'Neurotransmitters and peptides: whispered secrets and public announcements'. *Journal of Physiology*, 586, 5625–32 (2008). Van den Pol, A. N. 'Neuropeptide transmission in brain circuits'. *Neuron*, 76, 98–115 (2012).
20. Greenfield, S. A. & Collins, T. F. T. 'A neuroscientific approach to consciousness'. *Progress in Brain Research*, 150, 11–23 (2005).

9. TOMORROW

1. Although there is no retrospective time perception in sleep in general, chronaesthesia (the perception of ongoing time), seems to operate in dreams. William Dement conducted two studies which demonstrated that dream time was similar to real time. Because dreamers' eyes move under their eyelids very rapidly while they are dreaming, Dement was able to monitor sleepers and record the length of their dreams by observing their rapid eye movement. After recording this information, Dement would wake dreamers and have them write down a description of their most recent dream. He assumed that longer dreams would take more words to describe than shorter ones. When he compared the number of words in each dream report with the number of minutes the dream had lasted, he found that the longer the dream, the more words the dreamer used to describe it. In a related experiment, Dement woke sleepers while they were dreaming and asked them how long they perceived their most recent dream to have taken. Eighty-three per cent of the time they perceived correctly whether their dreams had been going on for a long time or a short time. With these experiments, Dement concluded that time in dreams is nearly identical to time in waking life. Time perception also seems to be the same in lucid dreaming. Then

again, perception of time as its actual passing, which appears to be what is being measured here, is not the same as the seeming retrospective absence of time between falling asleep and waking up with which we are all familiar. Dement, W. C. 'History of sleep medicine'. *Neurologic Clinics*, 23, 945–65, v (2005). Bolz, B. 'How time passes in dreams. (2009). http://indianapublicmedia.org/amomentofscience/time-passes-dreams/. Erlacher, D. et al. 'Time for actions in lucid dreams: effects of task modality, length and complexity'. *Frontiers in Psychology*, 4, 1013 (2013).

2. Taylor, B. N. *The International System of Units (SI)*. (NIST, 2001).
3. Hughes, D. O. & Tautman, T. R. *Time: Histories and Ethnologies*. (Michigan University Press, 1995).
4. Gell, A. *The Anthropology of Time: Cultural Constructions of Temporal Maps and Images* (Berg, 1992).
5. Trigg, G. *Encyclopedia of Applied Physics*. (Wiley, 2004).
6. Grondin, S. 'Timing and time perception: a review of recent behavioral and neuroscience findings and theoretical directions'. *Attention, Perception and Psychophysics*, 72, 561–82 (2010).
7. Tulving, E. 'Episodic memory: from mind to brain'. *Annual Review of Psychology*, 53, 1–25 (2002).
8. Originally published anonymously in *The Alternative: A Study in Psychology*. (London: Macmillan and Co., 1882).
9. Alexander, I., Cowey, A. & Walsh, V. 'The right parietal cortex and time perception: back to Critchley and the Zeitraffer phenomenon'. *Cognitive Neuropsychology*, 22, 306–15 (2005); see also Koch, G. et al. 'Underestimation of time perception after repetitive transcranial magnetic stimulation'. *Neurology*, 60, 1844–6 (2003); and Danckert, J. et al. 'Neglected time: impaired temporal perception of multisecond intervals in unilateral neglect'. *Journal of Cognitive Neuroscience*, 19, 1706–20 (2007).
10. Walsh, V. 'A theory of magnitude: common cortical metrics of time, space and quantity'. *Trends in Cognitive Science*, 7, 483–8 (2003); see also Bueti, D., Bahrami, B. & Walsh, V. 'Sensory and association cortex in time perception'. *Journal of Cognitive Neuroscience*, 20, 1054–62 (2008).
11. Tregellas, J. R. et al. 'Effect of task difficulty on the functional anatomy of temporal processing'. *NeuroImage*, 32, 307–15 (2006); see also Lewis, P. A. & Miall, R. C. 'A right hemispheric prefrontal system for cognitive time measurement'. *Behavioural Processes*, 71, 226–34 (2006); and Penney, T. B. & Vaitlingham, L. 'Imaging time' in *Psychology of*

Time (ed. Grondin, S.) 261–94 (Binglet UK Emerald Group, 2008); and Macar, F. & Vidal, F. 'Timing processes: an outline of behavioural and neural indices not systematically considered in timing models'. *Canadian Journal of Experimental Psychology*, 63, 227–39 (2009).

12. Mackintosh, N. J. *Animal Learning and Cognition.* (Academic Press, 1994); see also Jaldow, E. J., Oakley, D. A. & Davey, G. C. L. 'Performance of decorticated rats on fixed interval and fixed time schedules'. *European Journal of Neuroscience*, 1, 461–70 (1989).
13. Hugo Merchant, Deborah Harrington and Warren Meck at the Instituto de Neurobiología in Mexico have used single-cell recording studies in monkeys to identify complex neuronal networks involved in time perception, and have highlighted a crucial circuit connecting the cortex-thalamus-basal ganglia. They suggest that such a 'temporal hub' with a distributed network can account for the abstract properties of interval tuning as well as temporal illusions and inter-sensory timing, and that the interconnections built into this core timing mechanism in the brain are designed to provide a form of redundancy as protection against injury, disease or age-related decline. Merchant, H., Harrington, D. L. & Meck, W. H. 'Neural basis of the perception and estimation of time'. *Annual Review of Neuroscience*, 36, 313–36 (2013).
14. Grondin, 2010; see also Eagleman, D. M. 'Human time perception and its illusions'. *Current Opinion in Neurobiology*, 18, 131–6 (2008).
15. Eagleman, 2008.
16. Morrone, M. C., Ross, J. & Burr, D. 'Saccadic eye movements cause compression of time as well as space'. *Nature and Neuroscience*, 8, 950–4 (2005).
17. New, J. J. & Scholl, B. J. 'Subjective time dilation: spatially local, object-based, or a global visual experience?' *Journal of Vision*, 9, 4, 1–11 (2009).
18. Stetson, C., Fiesta, M. P. & Eagleman, D. M. 'Does time really slow down during a frightening event?' *PLoS One*, 2, e1295 (2007).
19. Tang, Y.-Y., Hölzel, B. K. & Posner, M. I. 'The neuroscience of mindfulness meditation'. *Nature Reviews Neuroscience*, 16, 213–25 (2015).
20. Eagleman, 2008.
21. One answer could be that some kind of low-level adaptation occurs: for example, adaptation to a flickering stimulus can lead to distortions in the estimate of the duration of subsequent stimuli, but only when presented in small, local regions of space. Since this effect was so localized, it implies there is some neuronal mechanism for timing in the more basic, early-stage visual areas. Yet another, completely different

idea would be at the other end of the processing scale within the brain, where duration is an index of the net amount of energy expended by neurons. Yet these are all still physical, objective events that would need to be appreciated ... but by what and where? Johnston, A., Arnold, D. H. & Nishida, S. 'Spatially localized distortions of event time'. *Current Biology*, 16, 472–9 (2006). Pariyadath, V. & Eagleman, D. 'The effect of predictability on subjective duration'. *PLoS One*, 2, e1264 (2007).

22. New & Scholl, 2009.
23. Stetson, Fiesta & Eagleman, 2007.
24. Matthews, W. J., Stewart, N. & Wearden, J. H. 'Stimulus intensity and the perception of duration'. *Journal of Experimental Psychology. Human Perception and Performance*, 37, 303–13 (2011).
25. Droit-Volet, S., Fayolle, S. L. & Gil, S. 'Emotion and time perception: effects of film-induced mood'. *Frontiers in Integrated Neuroscience*, 5, 33 (2011).
26. Levy, F. & Swanson, J. M. 'Timing, space and ADHD: the dopamine theory revisited'. *Australian and New Zealand Journal of Psychiatry*, 35, 504–11 (2001).
27. Parsons, B. D. et al. 'Lengthened temporal integration in schizophrenia'. *Neuropsychologia*, 51, 372–6 (2013).
28. Franck, N. et al. 'Altered subjective time of events in schizophrenia'. *Journal of Nervous and Mental Disease*, 193, 350–3 (2005).
29. Wittmann, M. et al. 'Impaired time perception and motor timing in stimulant-dependent subjects'. *Drug and Alcohol Dependence*, 90, 183–92 (2007); see also Cheng, R.-K., MacDonald, C. J. & Meck, W. H. 'Differential effects of cocaine and ketamine on time estimation: implications for neurobiological models of interval timing'. *Pharmacology, Biochemistry and Behavior*, 85, 114–22 (2006).
30. Tinklenberg, J. R., Roth, W. T. & Kopell, B. S. 'Marijuana and ethanol: differential effects on time perception, heart rate and subjective response. *Psychopharmacology (Berl)*, 49, 275–9 (1976).
31. Arzy, S., Molnar-Szakacs, I. & Blanke, O. 'Self in time: imagined self-location influences neural activity related to mental time travel'. *Journal of Neuroscience*, 28, 6502–7 (2008).
32. Kolb, B. et al. 'Experience and the developing prefrontal cortex'. *Proceedings of the National Academy of Sciences, USA*, 109 (suppl.), 17186–93 (2012).
33. Cooper, B. B. 'The science of time perception: stop it slipping away by doing new things'. *Buffer Blog* (2013).

34. Brown, J. 'Motion expands perceived time. On time perception in visual movement fields'. *Psychologische Forschung*, 14, 233–48 (1931).
35. Eagleman, 2008; see also Schiffman, H. R. & Bobko, D. J. 'Effects of stimulus complexity on the perception of brief temporal intervals'. *Journal of Experimental Psychology*, 103, 156–9 (1974).
36. Lucentini, J. 'It's neuron time'. *Science*, 17, 32–3 (Nov. 2003).
37. A space-time manifold is a mathematical model that describes space and time in a single continuum on scales ranging from subatomic to supergalactic, where time would be a fourth dimension, called a Minkowski space, complementing the familiar three dimensions of space. Such a manifold would be 'delocalized', in that it is not defined entirely by spatial coordinates.
38. Critchley, M. *The Parietal Lobes*. (Edwin Arnold, 1953).
39. Walsh, 2003.
40. Mitchell, C. T. & Davis, R. 'The perception of time in scale model environments'. *Perception*, 16, 5–16 (1987).
41. DeLong, A. J. 'Phenomenological space-time: toward an experiential relativity'. *Science*, 213, 681–3 (1981).
42. Rudd, M., Vohs, K. D. & Aaker, J. 'Awe expands people's perception of time, alters decision making and enhances well-being'. *Psychological Science*, 23, 1130–6 (2012).
43. Stetson, Fiesta & Eagleman, 2007.
44. Sergent, C., Baillet, S. & Dehaene, S. 'Timing of the brain events underlying access to consciousness during the attentional blink'. *Nature and Neuroscience*, 8, 1391–1400 (2005); see also Libet, B. *Mind Time: The Temporal Factor in Consciousness*. (Harvard University Press, 2004).
45. Synaptic connectivity in the mammalian cortex decays within a distance of 50um to 200um and wouldn't be affected by the difference in parameters of any single stimulus. Petreanu, L. et al. 'The subcellular organization of neocortical excitatory connections'. *Nature*, 457, 1142–5 (2009). Romand, S. et al. 'Morphological development of thick-tufted layer v pyramidal cells in the rat somatosensory cortex'. *Frontiers in Neuroanatomy*, 5, 5 (2011). Perin, R., Berger, T. K. & Markram, H. 'A synaptic organizing principle for cortical neuronal groups'. *Proceedings of the National Academy of Sciences of the United States of America*, 108, 5419–24 (2011). Boudkkazi, S., Fronzaroli-Molinieres, L. & Debanne, D. 'Presynaptic action potential waveform determines cortical synaptic latency'. *Journal of Physiology*, 589, 1117–31 (2011).

46. Chakraborty, S., Sandberg, A. & Greenfield, S. A. 'Differential dynamics of transient neuronal assemblies in visual compared to auditory cortex'. *Experimental Brain Research*, 182, 491–8 (2007).
47. However, the slow spread of activity within an assembly, over several hundred milliseconds in some cases, could simply be attributable to the additive effect of a very large number of synaptic connections, whereby the cumbersome process of neurotransmission will impede the signal speed enormously. Since the release, diffusion and action of a transmitter at its receptor takes approximately 0.75ms, couldn't this potentially lengthy time frame simply represent the summation of hundreds of sequential synaptic connections? Not really: even the timescale for a slow excitatory potential which could precede or result from traditional synaptic activity would last for a maximum of only 20msec. This would mean that the initial locus of the original stimulation in the centre of the assembly would most likely have decayed easily within the first hundred milliseconds or so – and, before then, the assembly would have the greatest activity at its perimeter: in fact, the very opposite spatial pattern is consistently seen, with the highest activity always in the centre.
48. Hestrin, S., Sah, P. & Nicoll, R. A. 'Mechanisms generating the time course of dual component excitatory synaptic currents recorded in hippocampal slices'. *Neuron*, 5, 247–53 (1990). Salin, P. A. & Prince, D. A. 'Spontaneous GABAA receptor-mediated inhibitory currents in adult rat somatosensory cortex'. *Journal of Neurophysiology*, 75, 1573–88 (1996). Katz, B. & Miledi, R. 'The measurement of synaptic delay, and the time course of acetylcholine release at the neuromuscular junction'. *Proceedings of the Royal Society of London B. Biological Sciences*, 161, 483–95 (1965). Sayer, R. J., Friedlander, M. J. & Redman, S. J. 'The time course and amplitude of EPSPs evoked at synapses between pairs of CA3/CA1 neurons in the hippocampal slice'. *Journal of Neuroscience*, 10, 826–36 (1990).
49. While we can never exclude completely the fact that conduction velocity is some four times slower with certain intra-cortical circuits, such evidence as there is would suggest that, if anything, it is up to ten times faster. Salami, M. et al. 'Change of conduction velocity by regional myelination yields constant latency irrespective of distance between thalamus and cortex'. *Proceedings of the National Academy of Sciences of the United States of America*, 100, 6174–9 (2003).
50. Taber, K. H. & Hurley, R. A. 'Volume transmission in the brain: beyond the synapse'. *Journal of Neuropsychiatry and Clinical Neuro-*

science, 26, iv, 1–4 (2014); see also Agnati, L. F. et al. 'Information handling by the brain: proposal of a new "paradigm" involving the roamer type of volume transmission and the tunneling nanotube type of wiring transmission'. *Journal of Neural Transmission* (2014).

51. Cheramy, A., Leviel, V. & Glowinski, J. 'Dendritic release of dopamine in the substantia nigra'. *Nature*, 289, 537–42 (1981). Greenfield, S. et al. '*In vivo* release of acetylcholinesterase in cat substantia nigra and caudate nucleus'. *Nature*, 284, 355–7 (1980). Greenfield, S. A. 'The significance of dendritic release of transmitter and protein in the substantia nigra'. *Neurochemistry International*, 7, 887–901 (1985). Nedergaard, S., Bolam, J. P. & Greenfield, S. A. 'Facilitation of a dendritic calcium conductance by 5-hydroxytryptamine in the substantia nigra'. *Nature*, 333, 174–7 (1988). Chen, B. T. et al. 'Differential calcium dependence of axonal versus somatodendritic dopamine release, with characteristics of both in the ventral tegmental area'. *Frontiers in Systems Neuroscience*, 5, 39 (2011). Mercer, L., del Fiacco, M. & Cuello, A. C. 'The smooth endoplasmic reticulum as a possible storage site for dendritic dopamine in substantia nigra neurones'. *Experientia*, 35, 101–3 (1979).
52. The conduction velocity in the cortex is between 0.1 and 0.5m/s for mammals; that is, for an action potential to travel along an axon, 100–500mm/sec (0.1–0.5mm/msec). Consistent with this speed, when stimulated from a source (for example, the thalamus, one synapse and a distance of 2.5mm away), the first sign of an assembly forming in the cortex is 5msec (Figure 10). But, subsequently, once the synaptic transmission between the two regions has occurred, the assembly takes up to an additional 15–20msec to reach maximum, which is astonishingly slow – too slow for synaptic transmission on its own – and yet too fast for only passive diffusion. González-Burgos, G., Barrionuevo, G. & Lewis, D. A. 'Horizontal synaptic connections in monkey prefrontal cortex: an *in vitro* electrophysiological study'. *Cerebral Cortex*, 10, 82–92 (2000). Stuart, G., Schiller, J. & Sakmann, B. 'Action potential initiation and propagation in rat neocortical pyramidal neurons'. *Journal of Physiology*, 505 (Pt 3) 617–32 (1997).
53. Gap junctions consist of the juxtaposition from each of two cells of two half-channels, each formed by proteins (connexins).
54. Draguhn, A. et al. 'Electrical coupling underlies high-frequency oscillations in the hippocampus *in vitro*'. *Nature*, 394, 189–92 (1998).
55. How comparable in terms of this net 'high' activity would be 200Hz oscillations with the signal intensity seen in assemblies? A frequency of

this speed would mean that an action potential is generated every 5ms, so about sixty in an assembly time window: our current calculations could account for this level of activity, which would show in the fluorescence signal as a periodic hyperpolarization via the electromagnetic mechanism, as there is no limit to how fast or slow the self-substance wave can oscillate. Huang H. et al. 'Remote control of ion channels and neurones through magnetic-field heating of nanoparticles', *Nature Nanotechnology*, 5, 602 (2011). Anastassiou C. A. et al. 'The effect of spatially inhomogeneous extracellular electric fields on neurones'. *Journal of Neuroscience*, 30, 1925 (2010).

56. Masson, G. S. & Ilg, U. W. *Dynamics of Visual Motion Processing: Neuronal, Behavioral and Computational Approaches*. (Springer, 2010).
57. For example, the expression of gap junctions in the retina is in inverse proportion to the availability of the transmitter dopamine, and it has been suggested that electrical junctions and chemical synapses usually work as a coordinated unit. He, S., Weiler, R. & Vaney, D. I. 'Endogenous dopaminergic regulation of horizontal cell coupling in the mammalian retina'. *Journal of Comparative Neurology*, 418, 33–40 (2000). Pereda, A. E. 'Electrical synapses and their functional interactions with chemical synapses'. *Nature Reviews Neuroscience*, 15, 250–63 (2014).
58. Chakraborty, Sandberg & Greenfield, 2007.
59. Bachmann, T. *Microgenetic Approach to the Conscious Mind*. (John Benjamins, 2000).
60. Vogel, E. K., Luck, S. J. & Shapiro, K. L. 'Electrophysiological evidence for a postperceptual locus of suppression during the attentional blink'. *Journal of Experimental Psychology. Human Perception and Performance*, 24, 1656–74 (1998); see also Sergent, Baillet & Dehaene, 2005.
61. Collins, T. F. T. et al. 'Dynamics of neuronal assemblies are modulated by anaesthetics but not analgesics'. *European Journal of Anaesthesiology*, 24, 609–14 (2007).
62. Chakraborty, Sandberg & Greenfield, 2007.
63. Kendig, J. J., Grossman, Y. & MacIver, M. B. 'Pressure reversal of anaesthesia: a synaptic mechanism'. *British Journal of Anaesthesia*, 60, 806–16 (1988).
64. Wlodarczyk, A., McMillan, P. F. & Greenfield, S. A. 'High pressure effects in anaesthesia and narcosis'. *Chemical Society Reviews*, 35, 890–8 (2006).

65. Wu, J.-Y., Xiaoying Huang & Chuan Zhang. 'Propagating waves of activity in the neocortex: what they are, what they do'. *Neuroscientist*, 14, 487–502 (2008); see also Muller, L. & Destexhe, A. 'Propagating waves in thalamus, cortex and the thalamocortical system: experiments and models'. *Journal of Physiology – Paris*, 106, 222–38 (2012).
66. Ferreira, P. G. *The State of the Universe*. (Phoenix, 2007).

Index

Notes are indicated by page number and note number: 226n45 indicates page 226, note 45